Microbiology

Bryan Larsen, PhD
Professor of Microbiology
and
Professor of Obstetrics and Gynecology
Marshall University School of Medicine
Huntington, West Virginia

Basic Science Monograph in Obstetrics and Gynecology

Council on Resident Education in Obstetrics and Gynecology
409 12th Street, SW, Washington, DC 20024-2188

Library of Congress Cataloging-in-Publication Data

Larsen, B. (Bryan).
Microbiology / Bryan Larsen.
p. cm. — (Basic science monograph in obstetrics and gynecology)
Includes bibliographical references and index.
ISBN 0-915473-14-3
1. Generative organs, Female—Microbiology. 2. Generative organs, Female—Infections.
I. Title. II. Series.
[DNLM: 1. Genital Diseases, Female—microbiology. 2. Infection. 3. Obstetrics.
WP 140 L334m]
QR171.F45L37 1991
618'.041—dc20 91-13382
CIP

ISBN 0-915473-14-3

1 2 3 4 5 / (9) 0 1 2 3 4

The development of this Basic Science Monograph on Microbiology was made possible by a grant from Parke-Davis.

Contents

Preface

The microscopic observation of living microorganims is at the heart of the development of microbiology, a science that is barely three centuries old. The work of the obstetrician and gynecologist has been with us much longer. However, it is noteworthy that the confluence of the two disciplines—microbiology and gynecology—occurred soon after microorganisms were discovered. Indeed, puerperal fever was among the first human diseases for which a microbial cause was recognized. Since this early liaison between these two disciplines, the science of each has evolved from gross and descriptive observations to an appreciation of molecular interactions that take place at the cellular level.

Although microbiology and obstetrics and gynecology developed separately, there have been acheivements along the way that show them continuing hand in hand. Since the introduction of aesepsis, which was born largely of an understanding of the etiology of childbed fever, the obstetrician-gynecologist has maintained an interest in microbiology. The need to find a therapy for syphilis led to Ehrlich's introduction of the concept of chemotherapy. Today antimicrobials are used constantly by gynecologists. The observations on rubella-associated birth defects led to an understanding of transplacental infection and ultimately to the use of vaccination. The development of immunology as an arm of microbiology permitted an understanding of maternal-fetal incompatibilities. Today, the cell culture techniques originally developed to study viruses are used for in vitro fertilization. Molecular genetics, which is currently so important in all branches of science, traces its history to observations of the role of DNA in transforming pneumococcal cells.

This monograph is intended to strengthen the connection between microbiology and obstetrics and gynecology by summarizing the basic concepts of microbiology that are central to understanding the infectious disease processes seen in female patients. This goal can only be reached by describing both the biology of microorganisms and the host's reaction to microorganisms. The true essence of infectious disease is neither the virulence of the microbe nor the susceptibility of the host, but the host-parasite interaction.

The chapters that follow emphasize the principles and mechanisms involved in the infectious diseases of current interest. The role and limitations of the diagnostic laboratory will also be described. Because it emphasizes basic mechanisms of disease and the source of the microorganisms involved, this book provides information that will be useful for both those infections we encounter today and those that will become problems in the future. The importance of being prepared to deal with emerging infectious disease problems is underscored by such diseases as toxic shock syndrome and the acquired immune deficiency syndrome (AIDS), which have reached prominence in the past decade.

Because it is not possible to include everything that now comes under the rubric of microbiology, there have been some conscious omissions from this book. This work will primarily emphasize the bacteria that are responsible for a very large proportion of the infectious disease problems of women. Fungal and parasitic infections will be noted as appropriate but will not receive exhaustive treatment. Virology has become a very large discipline, particularly as the molecular biology of viral replication and immunity has come to the fore. Consequently, the subject of viral infections requires a treatise of its own. However, the viral-infection issues that are currently most pressing will be included.

Because the purpose of this work is to ensure a basic fund of microbiological knowlege for the obstetrican-gynecologist, little effort will be expended on providing a "cookbook" for therapy. There are many sources of therapeutic protocols, but these are only rational in light of the fundamentals of infection and immunity. Therapy is discussed from the viewpoint that sound use of antimicrobial drugs stems from a clear understanding of the principles of infection. This type of knowledge will be most adaptable to the ever-increasing number of available therapeutic agents.

To anyone who believes that microbiology has no connection to clinical obstetrics and gynecology, this book will demonstrate that the two fields share many points of confluence. This work will emphasize the relevance of microbiology to the well-being of the female patient. The fundamental knowledge of infectious agents possessed by the obstetrician-gynecologist will help him or her to provide the best care for female patients and to enjoy the profound satisfaction that comes from an appreciation of the fundamental mechanisms involved in health and disease.

B.L.

Microbiology

Part I

The Biologic Basis for Obstetric and Gynecologic Infectious Disease

Gentlemen, that you may get satisfactory results with this sort of treatment, you must be able to see with your mental eye the septic ferments as distinctly as we see flies or other insects with the corporeal eye. If you can really see them in this distinct way with your intellectual eye, you can be properly on your guard against them; if you do not see them you will be constantly liable to relax your precautions.—Lister

1

Overview of the Biology of Microorganisms

INTERACTIONS OF MICROBES WITH THEIR ENVIRONMENTS

During basic science training, each medical student is required to learn microbiology. As a result, he or she gains an appreciation of the biology of microorganisms and is expected to see these organisms as living cells with many interesting biologic and biochemical capabilities. To the practicing physician, the term "microorganism" usually conjures the image of an infectious agent and almost as quickly a readiness to mentally scan the list of available antimicrobial agents to select one suitable for destroying the infectious organism. But this is not the usual context for most microorganisms. It should be remembered that the vast majority of microorganisms are not pathogenic for humans. Perhaps equally important, it should be emphasized that many organisms that associate with the host tissues are basically benign.

Although the reader has undoubtedly had some formal exposure to the discipline of microbiology, it will be useful to review some of the most important characteristics of microorganisms before discussing the distinguishing features of that special class of microbes that cause disease. The biology of microorganisms has direct relevance to infectious diseases.

The term "microorganism" refers to any organism too small to be seen with the naked eye and as a result encompasses a diverse group of life forms, including bacteria, viruses, fungi, and protozoans. Of the different life forms, the greatest emphasis will be placed on the bacteria, which cause a significant proportion of the treatable infectious diseases that the obstetrician-gynecologist must face.

Bacteria are ubiquitous in the environment and on the mucosal surfaces of animal hosts. These microorganisms are able to obtain the necessities for survival within the environments they colonize. To appreciate the fact that only approximately 100 of the thousands of microbial species have specific relevance to human health, it is appropriate to note the various relationships microorganisms may have with their environment. Figure 1-1 summarizes these relationships.

Among the possible relationships between microorganisms and a living host, the two extremes are complete antagonism of the host toward the microorganism and overwhelming antagonism of the organism toward the host. The former is exemplified

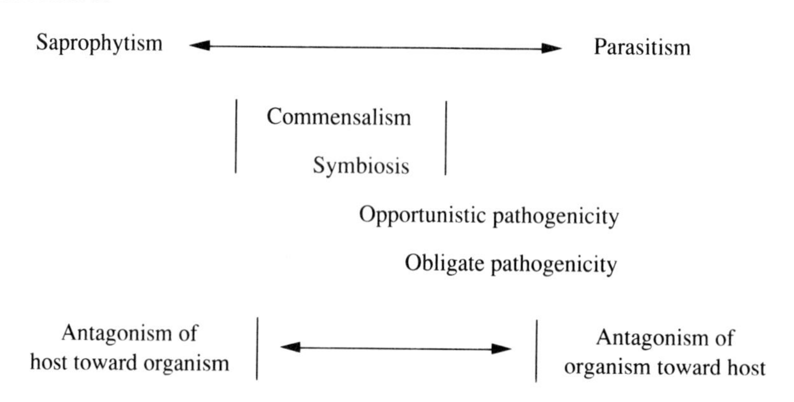

Fig. 1-1. *Possible relationships between microorganisms and a living host.*

by microorganisms called saprophytes. These organisms grow on nonviable materials and virtually never associate with an animal host either as a cause of disease or as normal flora. Saprophytic organisms are generally of little interest to physicians. However, in severely compromised hosts, organisms that have been traditionally considered merely as saprophytes, for example *Bacillus* spp., can cause infection, which blurs somewhat the distinctions between saprophyte and parasite.

The other extreme is exemplified by a microorganism that always causes disease when it associates with the animal host. Such organisms have the ability both to invade the host and to cause damage. Consequently, the term "pathogen" is appropriately applied to such organisms.

Most microorganisms that associate with human hosts do not fit neatly into the category of frank pathogen. Therefore, a clear understanding of microorganisms should not include any attempt to characterize all microorganisms as pathogens or nonpathogens. Figure 1-1 illustrates several bacterial interactions with the host that lie between the extremes of obligate parasitism and saprophytism. In the theoretic center, there may be a balanced relationship between the host and microorganism in which neither harms the other. If the microorganism requires the host for its survival and the host in turn depends on the microorganism for some vital function, the relationship may be considered to be symbiotic. If no harm comes to either party from the relationship, but the relationship is not essential to either host or microbe, the condition may be considered one of mutualism or commensalism.

Microorganisms commonly associate with the host in a mildly antagonistic or parasitic manner that is exploited at times when the host is immunocompromised. Such organisms are best described as opportunists.

It should be clear from this representation that microorganisms are difficult to categorize as pathogenic or nonpathogenic and that the space between these extremes is a continuum of host-microbe interactions.

ANATOMY OF THE BACTERIAL CELL

Cytoplasmic Constituents

The limiting membrane of the bacterial cell contains most of the structures that are involved in the metabolic and reproductive functioning of the bacterial cell. Bacteria are prokaryotic organisms, which means that their DNA is not organized into a membrane-bounded nucleus. The eukaryotes have membrane-bound nuclei and other intracellular organelles. Within the cytoplasm are many of the proteins that carry out the metabolic activities of the cell, although some enzymatic processes are associated with the cell membrane. The ribosomes, which are complex organizations of RNA and protein, are the site of protein synthesis. Each of these structures and metabolic functions represent sites that may be attacked by antimicrobial drugs, and hence have relevance to the study of infectious disease.

The Cell Envelope

The outer boundary of bacterial cells is complex and differs with the gram reaction of the cell wall. Figure 1-2 schematically depicts the elements that comprise the various structures that form the outermost portions of the bacterial cell.

The cytoplasm of bacteria is bounded by a typical biologic membrane called the plasma membrane or cytoplasmic membrane. This membrane is not simply a structure that corrals the intracellular soup, but is actively involved in the uptake of nutrients from the environment. Some nutrients are able simply to move across the membrane to achieve cytoplasmic concentrations comparable to external concentrations. However, active

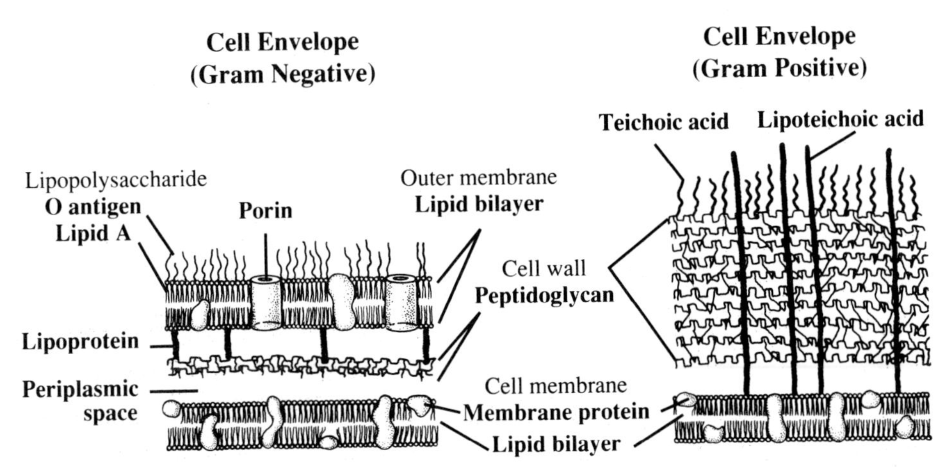

Fig. 1-2. *Composition of the structures that form the outermost portions of the bacterial cell.*

transport mechanisms that require energy expenditure may concentrate some compounds in the cell at levels far exceeding the extracellular concentration. Enzymes that transport the substrate across the membrane are called permeases.

Newly synthesized proteins destined for export to the extracellular or periplasmic environment are also transported across the plasma membrane by energy-requiring reactions. The molecules to be transported are marked by signal sequences represented by 20-40 amino acids at the N terminus. These signal sequences are predominantly on the hydrophobic portion of the peptide, which allows appropriate interaction with the plasma membrane. The signal peptide is removed after protein translocation.

Another activity that occurs at the cell membrane is oxidative metabolism. The enzymes involved in this process are organized in precise spatial relationships, and their attachment to the membrane allows the appropriate geometry required for the oxidoreductase enzymes to operate in appropriate sequence.

The cytoplasmic membrane forms an anchorage site for cellular DNA, which after replication must be segregated into each of the daughter cells. A significant number of the ribosomes also affiliate themselves with the cytoplasmic membrane.

Outside the plasma membrane lies the cell wall, which is the main distinguishing feature between gram-positive and gram-negative bacteria. The cell wall is in both cases a rigid polymeric structure consisting of peptidoglycan (which is a mesh-like copolymer of *N*-acetyl glucosamine and *N*-acetylmuramic acids) cross-linked with small peptide bridges. There are some differences in the exact structure of the peptidoglycan among species, and the quantity and substituents on the cell wall material differ markedly depending on the Gram-stain reaction of the cell.

Gram-positive cells have thick layers of peptidoglycan, whereas gram-negative cells have approximately 1/10th the thickness of this material. The gram-positive cell wall material may have teichoic acid (a ribitol-glycerol copolymer of variable molecular structure, linked by phosphodiester bonds) associated with it. Teichuronic acids (glucosidically linked copolymers of *N*-acetyl galactosamine and glucuronic acid) may be present with or in lieu of teichoic acids. In addition, lipoteichoic acid molecules may extend from anchorage points in the cell membrane outward through the cell wall material and form, along with the teichoic acids of the cell wall, the major surface antigens of the gram-positive bacterial cell. The wall will usually carry complex polysaccharides that are highly specific for the microbial species.

The much thinner cell wall layer of the gram-negative cells has an exceedingly complex structure overlaying it. A periplasmic space exists between the cell membrane and the base of the cell wall. This space is the site of accumulation of materials in transit from the outside to the inside of the cell, or the reverse. Enzymatic activities present in this space have relevance for the therapy of gram-negative infections. The gram-negative cell is unique in that it has an outer membrane attached to the exterior of the cell wall by means of lipoproteins. Thus, the gram-negative cell has both an inner and an outer membrane. Consequently, its surface structure is markedly different than that of gram-positive cells.

The outer membrane of the gram-negative cell is a bilaminar leaflet, with the outer layer composed of lipopolysaccharide or endotoxin. This membrane also contains porins, which are multimeric protein complexes that form pores through the membrane. These pores serve as channels for small ions and may also be the site of bacteriophage attachment. The outer surface of the gram-negative cell also contains a variety of protein components that function as adhesive factors or interact with host defense factors in ways that will be described in a later chapter.

Additional components of bacterial cells that lie outside the cell wall are flagella. Flagella are relatively long, proteinaceous structures that are the organs of motility in gram-positive bacteria. Bacteria may contain pili or fimbriae, which are appendages that have a hairlike appearance upon magnification by electron microscopy. The normally anionic bacterial surface would tend to be repelled by the anionic cell surface of the host cell. These repulsive forces may be overcome by fimbriae.

Many bacteria also possess a layer of loosely organized complex carbohydrate described as the capsule. As discussed in chapter 2, the capsule plays a particularly important role in protecting bacteria from phagocytosis and consequently represents one of the virulence attributes of microorganisms.

ANATOMY OF THE EUKARYOTIC CELL

Fewer eukaryotic microorganisms than bacterial species are of relevance to the female genital tract. An extensive description of the anatomy of these organisms will not be undertaken. As implied by the designation "eukaryotic," these organisms are distinguished from bacteria by the presence of a true nucleus bounded by a nuclear membrane. They also contain membrane-bounded intracellular organelles such as mitochondria. The fungi are enclosed by a cell wall made of a complex network of mannans, whereas the protozoans do not have a cell wall but have only the plasma membrane exposed to the extracellular environment.

ANATOMY OF VIRUSES

Mature virus particles, free from the intracellular environment of host cells, are relatively simple since they do not possess the metabolic structures and functions characteristic of bacteria or eukaryotic cells. The complete virus particle, or virion, consists of genetic material that may be DNA or RNA enclosed in a protein coat. The coat of the virus particle consists of only a few proteins. If the virus exits its host by budding through the surface of the cell, it will also have some host cell membrane elements and some virus-specified elements. In the case of those viral species that have RNA as the genetic material, the mature virion must carry an enzyme that is able to transcribe RNA since host cells do not possess such enzymes. The RNA virus either carries an RNA replicase, which produces RNA transcripts from an RNA template, or reverse transcriptase, which generates DNA transcripts from RNA templates in the case of the retroviruses. In contrast to bacteria, fungi, and protozoa, which carry thousands of proteins, the mature virion consists of only a few macromolecular components.

PHYSIOLOGY OF MICROORGANISMS

As part of their survival strategy, microorganisms appear to have the characteristics of rapid growth and metabolism. With the exception of bacterial sporulation, which occurs in relatively few bacteria of medical importance, dormancy is rarely part of the microorganism's way of life. Growth of microorganisms is limited, with the size of most organisms remaining fairly constant. Rather than increasing in size, bacteria increase in numbers. The physiologic activities of bacteria, therefore, include intake of nutrients, energy production, biosynthesis of structural elements, and replication of DNA followed by cell division.

It is appropriate to provide a brief summary of these activities because they are relevant to microbial growth in human hosts, disease production, antimicrobial chemotherapy, and methods of identification,

The microorganism must produce energy from the nutrients in its environment, whether that environment is a test tube filled with artificial growth medium or a serosanguinous collection above the vaginal cuff produced during a hysterectomy. Microorganisms produce energy from a variety of substances. However, all of the methods of energy production require that the energy of electrons that form chemical bonds be transferred to a useful form. The reducing power required for the energy needs of bacteria is in the form of electrons. These electrons, of course, are not free, but are carried by hydrogen atoms. Substrates that provide energy are oxidized by the loss of electrons, which are passed in stepwise fashion to electron acceptors. The electrons are ultimately transferred to compounds that can provide this energy for many different reactions within the cell. Thus, ATP or the reduced forms of NAD or NADP are the "coin of the realm" for energy-requiring processes within the cell.

In generating energy, the cell must have not only a source of electrons but also a final repository for them after the energy has been obtained. Two types of metabolism used by medically important bacteria are distinguished from each other by the final electron acceptor. If oxygen is used as the final electron acceptor, the electrons in the form of hydrogen atoms are added to molecular oxygen to form water. This process is described as respiration and results in substantial ATP production. Of course, this process is aerobic, and it may be a preferred method of generating energy because of the large amount of energy that can be produced.

The second major method of producing energy is fermentation. This process does not employ oxygen as a final electron acceptor but rather splits a substrate and uses part as an electron donor and part as the final electron acceptor. In addition to the fact that fermentations are anaerobic processes, it should be noted that the products of the fermentations are organic acids and alcohols. These products may accumulate in areas where anaerobic bacterial growth is occurring and may be helpful in relation to diagnostic procedures.

In addition to substrates that can be used to produce energy, the nutritional requirements of microorganisms include carbon and nitrogen sources, which are used in the synthesis of structural components of the cell; nucleic acids; and enzymes required to catalyze the anabolic and catabolic reactions needed for cell growth. In the bacteria usually associated with human colonization or human infection, these needs are met by the proteins and carbohydrates present in the complex microenvironment of the host tissues. However, bacteria benefit from the ability to utilize carbohydrates other than glucose. It will be shown later that knowing the repertoire of carbohydrates that microorganisms are able to utilize can aid in differentiating various species.

To utilize proteins and complex carbohydrates as sources of carbon or nitrogen, bacteria produce hydrolytic enzymes that degrade these substances to manageable subunits. Likewise, enzymes that degrade polynucleotides may also be produced by bacteria. Some of the exoenzymes released into culture medium by microorganisms may be measured as one way to confirm the identity of a particular microorganism. In vivo, these enzymes not only may aid in the acquisition of nutrients, but in some cases they are involved as virulence factors.

A further nutritional requirement of bacteria is the vitamin cofactor needed for various metabolic reactions. Some bacteria are sufficiently versatile to synthesize their own vitamins, whereas others require preformed cofactors. Frequently, organisms that are adapted to growth in or on a mammalian host are accustomed to obtaining vitamins from the host. As a consequence, very rich media are generally used for the cultivation of medically significant microorganisms .

Iron and several other inorganic micronutrients are required for bacterial growth. Iron is apparently of paramount importance, both as an enzyme cofactor of several well-defined metabolic systems (including ferridoxins or superoxide dismutase) and in less well defined virulence-enhancing roles. The major problem with iron is its insolubility. In the host the problem is solved by such iron-carrying proteins as transferrin. Bacteria synthesize proteins that are able to bind iron to aid in its transport into the cell. These compounds are called siderophores. Siderophores are efficient chelators of iron and in some instances are able to attract it away from transferrin. *Neisseria gonorrhoeae* can obtain iron from transferrin and lactoferrin because it possesses surface receptors for these iron-bearing proteins.

Bacterial Growth

Bacteria utilize the available nutrients and replicate as quickly as the environmental conditions permit. Artificial media provide optimal conditions for the growth of bacteria. Under these conditions, after a period of adaptation to the medium, the population doubles at regular intervals and the numbers of viable microorganisms increase according to a geometric progression. When nutrients become depleted or toxic metabolites accumulate, the growth rate slows and eventually the number of viable cells becomes stationary. Ultimately, the bacteria begin to die and the number of viable cells declines. This type of growth is characteristic of a closed system in which the amount of nutrient is limited, the metabolic by-products remain in the culture medium, and all additional bacteria produced by cell division remain in the culture medium.

Within the tissues of an infected host, however, a different situation exists. The host is not a closed system. Microorganisms may be removed by host defense mechanisms, metabolic by-products may be absorbed or redistributed, and fresh nutrients may enter the infected site. In addition, there may be more than one species of microorganism at the site of infection. Such features of the infected host could have a profound effect on the way in which microorganisms grow in vivo.

Bacterial Genetics

The study of bacterial genetics has been developing rapidly since the discovery that the virulence of the pneumococcus resulted from the ability of DNA to alter the phenotype of this bacterium. Today, volumes are available, and these paragraphs will not presume to fully cover the topic. The basic point included here is that bacterial cells possess a single chromosome consisting of double-stranded DNA. The information contained in the bacterial chromosome is subject to the same processes that all organisms use, namely, the replication of the chromosome when the cell reproduces itself and the expression of the chromosome by the transcription of the genetic information to messenger RNA, which is then translated into proteins that have both structural and enzymatic activity.

The fact that only one chromosome has to be replicated for reproduction to occur obviates the need for the complex mitotic machinations that occur in eukaryotes. However, the bacterial chromosome—being a closed circular piece of DNA—does present some physical problems. One of the currently important aspects of this process is the unwinding of the DNA, which is required to allow the necessary replicative enzymes to gain access to the individual DNA strands. The twisted pair of DNA strands does not untwist along the full length of the chromosome. Under the action of topoisomerase enzymes, the double strand untwists at the site where the replicative activity will be carried out and supercoils elsewhere. The topoisomerase enzymes are affected by the quinolone class of antibiotics, as will be discussed in greater detail later.

The replication of the bacterial chromosome is a very efficient process. Bacteria must have a new copy of the chromosome each time they divide, which could be as frequently as every 20 minutes. The chromosome of *Escherichia coli* consists of five million base pairs, which means that synthesis of new DNA must occur at the rate of 4,000 bases per second. It has been suggested that the uncoiling process of the chromosome must occur at about 6,000 revolutions per minute. The efficiency of this system is difficult to envision. One of the reasons for such effective replication is the bidirectional DNA replication involving each strand. Furthermore, before one cycle of replication has been completed, bidirectional replication of the newly replicated region of the chromosome may begin.

The genes of the bacterial chromosome, like those of any other biologic system, can be altered by mutations, and changes in the bacterial phenotypic characteristic can result from deletions or insertions in the chromosome. The discovery that DNA can be cut and genetic material can be spliced in has been one of the most powerful findings of molecular biology. This is not an artificial process, but one that scientists have exploited by using the enzymes that take care of such details involved in replication of DNA as opening and closing DNA strands, repairing breaks, and replacing defective regions. All of these processes occur in the normal bacterial cell, despite the breakneck speed at which other processes involved in replication and gene expression occur.

In addition to the bacterial chromosome, genetic material that exists as autonomously replicating closed circles of DNA may be found in the cytoplasm. These circles of DNA are called plasmids. They may have profound effects on the nature of the cell since their gene products may be expressed, and since there may be hundreds of copies of these plasmids in the bacterial cell. As noted above, there are mechanisms that allow DNA strands to be opened and then closed again. These processes are essential to the insertion of new material into the bacterial chromosome. The DNA of plasmids is not always separate from the bacterial chromosome but may become incorporated into the bacterial genome.

At the heart of molecular biology as we know it today is the discovery of extrachromosomal pieces of genetic material within the cell, along with the finding that these pieces of DNA can be incorporated into the bacterial chromosome and that these genetic elements may be expressed whether they are incorporated into the bacterial chromosome or not. Bacteria are able to receive genetic material from other sources and use that genetic material as if it were part of their own genome. Specifically, a bacterial cell can under some circumstances take a naked piece of DNA into its cytoplasm and that DNA may be inserted into the bacterial chromosome by the recombination of the exogenous DNA with the bacterial DNA. This process of transformation (not to be confused with the transformation of normal cells to cancerous cells) was first demonstrated with the pneumococcus, which could be transformed from the avirulent unencapsulated form to the virulent encapsulated form by the acquisition of the genes that specify capsule formation. Not only does this demonstrate that microorganisms have the ability to adapt by expressing traits of newly acquired genes, but it also indicates that virulence may be affected by certain specific genes and their products.

The bacterial cell has two other means of exchanging genetic material: conjugation and transduction. The transfer of exogenous DNA by the process of conjugation is a sexual process involving the contact of two bacteria and the transport of donor DNA to the recipient cell. Conjugation occurs between closely related organisms, one of which must have the "fertility" factor. This factor allows it to produce a conjugation appendage, the sex pilus, which attaches to a recipient cell without the fertility factor. For convenience and perhaps because of our recognition of the universality of genetic exchanges, the two cell types are often referred to as male and female, respectively. This process has been shown to be responsible for some instances of bacteria acquiring genes that confer resistance to antibiotics. This is an important finding because it means that resistance to an antibiotic may spread to susceptible strains of bacteria, a fact which could have profound implications for the future of antimicrobial chemotherapy.

The second method of genetic exchange is transduction. In this method a bacterial virus (bacteriophage) acquires some bacterial DNA either by accidentally packaging bacterial DNA into some of the mature bacteriophage (generalized transduction) or by picking up some bacterial genes during the process of lysogeny (specialized transduction). Lysogeny is a process by which the virus, rather than simply replicating itself in the bacterial cytoplasm and producing new virus particles, inserts its genome into the bacterial chromosome. When the process of lysogeny ends with the virus genome disengaging from the bacterial chromosome (usually after many rounds of bacterial replication), the virus genome may carry along some of the bacterial genes and these become packaged into new virus particles. In each type of transduction, the mature virus attaches to an uninfected bacterial cell and introduces the viral DNA into the bacterial host. However, this viral DNA is not purely of viral origin and the bacterial genes accidentally carried by the virus may be expressed as new traits in the recipient bacterium.

This discussion has indicated the manner in which bacteria may acquire exogenous genetic material. Regardless of how that material arrived in the cell, we now are aware that small pieces of extrachromosomal DNA may exist within the cytoplasm of the cell. These are usually closed circles of double-stranded DNA that autonomously replicate. These genetic elements are referred to as plasmids, and recognition of their significance has increased as more knowledge of their function has become available. We now know that many of these plasmids confer virulence upon a bacterium. Others permit the bacterium to resist antibiotics. As already noted, the maleness of a bacterial cell and hence its ability to transfer DNA by conjugation is located on a plasmid. Indeed plasmids may be even more important than would be suggested by their prevalence in the cytoplasm. Some plasmids are able to become integrated into the bacterial chromosome, and it is very probable that many such cryptic plasmids exist. The special capabilities conferred upon a cell by plasmids may help the bacterium to survive adversities in its environment, whether the bacterium is free-living or is living within a mammalian host.

Basic Taxonomic Considerations

THE NEED FOR TAXONOMIC DESIGNATIONS

It may seem so obvious as not to bear mention, but microorganisms are incredibly diverse and do not all possess the same ability to cause infectious diseases in humans. Because of the profound virulence associated with some organisms, the ability to quickly identify these organisms and distinguish them from other organisms is a medical imperative. One of the reasons why microorganisms are identified is to aid in selecting appropriate therapy. The susceptibility of certain bacteria to antimicrobial drugs is quite predictable in some cases, and when such an organism is identified, therapy can be instituted without testing the antimicrobial susceptibility pattern of the organism. In the case of those organisms with a typically unpredictable susceptibility pattern, antimicrobial testing may be initiated immediately by the laboratory without further consultation with the physician.

The degree to which identification of an infectious agent is pursued differs according to the needs of the patient and the physician. Generally, to adequately document the etiology of the disease, identification to the species level is desirable. However, adequate therapy may be instituted in many cases simply based on a genus-level identification. In the case of epidemiologic investigations, it may be necessary to go beyond speciation to determine whether a certain strain of microorganism is responsible for multiple cases of infection.

THE GENETICS OF BACTERIAL TAXONOMY

The basic genetic endowment of any organism distinguishes it from other similar organisms. Ultimately, the distinguishing characteristics of microorganisms arise from the specific genes present on the bacterial chromosome and the expression of these genes with the given environmental conditions. The specific arrangement of the nucleic acids on the bacterial chromosome is too vast to determine, but the G-C ratio—the proportion of the chromosome composed of guanosine-cytosine pairs—provides a crude look at the entire bacterial chromosome. The relatedness of microorganisms has been based on this relatively easily measured characteristic. In the future, the presence of specific genetic sequences will become the basis for diagnostic tests. Diagnostic gene probes are being developed for a variety of bacteria and viruses, although at this writing the only test approved by the Food and Drug Administration is for human papillomavirus.

The classical methods of identifying microorganisms depended on the phenotypic characteristics of the microorganisms. What are the shape and morphology, the Gram stain reaction, and the fermentation capabilities of the organism? All of these are derived from the genomic composition. However, diagnostic laboratories are currently able to perform efficient and cost-effective phenotypic characterizations of microorganisms, whereas genotypic characterization will not be routinely practiced in laboratories for some time.

In noting that taxonomic differences among microorganisms arise from the genetic material of the cell, it is important to realize that not all of the genetic information is encoded on the bacterial chromosome. Extrachromosomal elements may contain genes that specify antibiotic resistance factors or other virulence factors. Some of the well-recognized virulence factors are specified by segments of the bacterial chromosome that originated as separate genetic elements that have been integrated into the bacterial chromosome. The genes or gene products of extrachromosomal elements may also be used for taxonomic characterization of microorganisms.

DIAGNOSTIC USES OF MICROBIAL CHARACTERISTICS

The Physician and the Laboratory

The physician's use of the diagnostic microbiology laboratory begins with the suspicion of infection in a patient. Even before involving the diagnostic microbiology laboratory, however, the physician can perform some tests to confirm his or her suspicions about the nature of the infection. After examination of the patient and identification of specific symptoms, the physician may wish to undertake some microscopic evaluations of the infected material. The physician usually diagnoses vaginal infections by direct microscopic examination. Dipstick urinalysis may be employed to give a preliminary indication of infection, and this may be followed by microscopic examination of the urine sediment. A Gram stain can be performed in a physician's office or on the ward in a hospital and will be useful if the site of infection does not contain a normal flora or if a specific microbial morphotype is suspected. A white blood cell count and differential may be obtained much more expeditiously than a bacterial culture, a fact that may influence the initial decision regarding the value of bacterial culture.

The physician's familiarity with the diagnostic capabilities of the laboratory will guide the subsequent decision about whether to use the laboratory's services. Certainly, bacterial cultures are more readily available than viral cultures in most institutions. Some viral diagnostic studies must be sent to remote laboratories. In cases of suspected viral infection when confirmation will not alter therapy, the physician might forego the option of viral culture.

The diversity of the infectious agents relevant to the obstetrician-gynecologist and the differences in the diagnostic capabilities of microbiology laboratories make it appropriate to discuss the things that are done in the diagnostic microbiology laboratory, particularly as these activities relate to the biology of the microorganisms as surveyed in this chapter.

Physiologic Features of Microorganisms

As indicated by the discussion of the physiology of microorganisms, bacteria are tremendously diverse with respect to the metabolic pathways used, the precise structure of various components of the cell, and ultimately the structure of the bacterial chromosome. All of these characteristics may be exploited by the clinical microbiology laboratory in fixing the identity of an organism isolated from a clinical infection. However, before any of these features are determined, it is necessary to obtain the offending organism in pure culture. Because different bacteria possess different individual characteristics, mixtures of organisms may obscure the individual features of each species present in the mixture.

The laboratory will first propagate the organisms present in the clinical specimen on a rich medium. As mentioned previously, many of the organisms capable of causing clinical infection are fastidious about growth requirements. In order not to miss fastidious organisms, highly enriched media are generally employed for primary isolation.

Once the organism or organisms are obtained in pure culture, some of the basic characteristics are determined, including the Gram stain reaction and microscopic morphology of the cell. These initial observations will guide the microbiologist in determining which physiologic characteristics of the organisms are relevant to their identification.

Although the specific tests conducted by the microbiology laboratory are outside the scope of this chapter, it is appropriate to describe some of the tests that are performed on bacteria. One category of tests involves the growth conditions of the organism, such as the ability to grow in the presence of oxygen or the ability to grow at an elevated temperature. It is possible to make microbiologic growth media containing only one carbon source. If a microorganism can grow on that medium, its ability to utilize that carbon source is demonstrated. Various sugars are often used in these tests, although complex carbohydrates such as starch may be tested. Testing for the oxidative or fermentative use of carbohydrates is also employed for certain organisms.

Another class of tests identifies specific products of microbial metabolism. These tests may simply be for the production of acid or gas from sugars or may involve the specific chromatographic identification of the organic acids produced. Simple tests for the presence of specific enzymes (DNase, cytochrome oxydase, or catalase, which breaks down hydrogen peroxide) are also useful in identifying bacteria.

Finally, the ability of organisms to resist the harmful action of inhibitors such as bile salts or certain antibiotics (e.g., bacitracin) is used to differentiate bacterial isolates. Together, all of the various physiologic features that may be tested can be used in logical schema to identify a microorganism. Such methods of identification apply mainly to bacteria. Because fungi and protozoal pathogens have rather striking microscopic morphology, microscopic examination is more useful for most infections involving these types of organisms. Viruses, of course, do not have a metabolism of their own and consequently have no physiologic features to examine.

Immunologic Features

All microorganisms display on their surfaces proteins and other complex compounds that can be recognized as antigens. These may be purified and injected into animals to raise antibodies. If the resulting antibodies do not cross-react with antigens of other microorganisms or if cross-reacting antibodies are assiduously removed by adsorption with cross-reacting organisms, the antiserum then represents a specific reagent that may be used to unequivocally identify a microorganism. This technique is applicable to bacteria isolated by culture or may be used to detect microbial antigens present on microorganisms obtained from an infected patient. More importantly, immunologic detection methods are applicable to organisms that are cumbersome to culture, such as viruses or chlamydia.

The practical use of antibodies raised against specific microbial antigens depends on an appropriate detection method. These methods include the agglutination test, the fluorescent antibody test, and the enzyme-linked immunosorbent assay (ELISA).

Agglutination may involve the addition of specific antibody to a suspension containing an unknown organism, in which flocculation of the organism indicates reaction with the antibody. The antibody may be linked to some other substance such as latex beads or charcoal particles. These particles then agglutinate with the specific antigen.

Fluorescent antibody tests involve labeling the antibody with a fluorescent dye. If the antibody reacts with its specific antigen, the antigen will fluoresce under a microscope with ultraviolet illumination. Such a test may be useful for detecting microbial antigens directly within clinical specimens. However, the drawback of this type of test is that it may require a fair amount of technical experience to obtain optimum results.

Enzyme immunoassay techniques may also be applied to samples in which the microbial enzyme remains in its biologic matrix. The antibody is affixed to a solid support such as the bottom of a plastic well. The sample that contains the antigen (microorganism) reacts with and binds to the antibody fixed to the plate. Unbound material is washed away, and additional antibody (this time tagged with an enzyme) is allowed to react with the antigen bound to the well. The unbound enzyme-tagged antibody is washed away, and the substrate for the enzyme is added. If the enzyme is present, a color reaction caused by enzymatic cleavage of the substrate occurs and the presence of antigen is proved. Figure 1-3 shows how the ELISA is used for antigen detection.

The immunologic techniques described thus far have been devised for the detection of the microbial antigen. However, in an infected individual an antibody response occurs, and this antibody response may also be exploited for diagnostic purposes. In this case, the

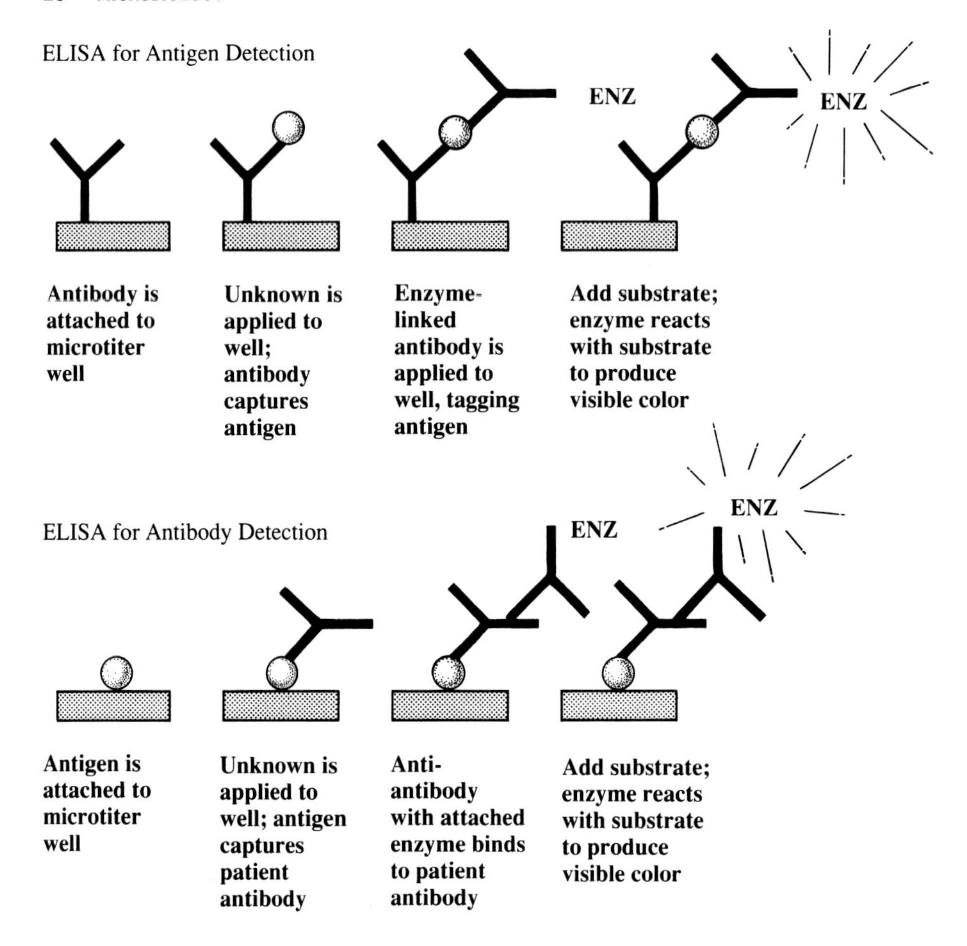

Fig. 1-3. *Steps involved in the use of the enzyme-linked immunosorbent assay (ELISA) for antigen and antibody detection.*

microbial antigen can be used as the reagent and the patient serum can be used as the unknown to be tested. The ELISA and other immunologic tests can be adapted for the detection of antibodies (see Fig. 1-3). In this arrangement the antigen, which may be a microorganism or a specific part of the organism, is attached to the well and reacted with the patient's serum. The antibody bound to the antigen is detected by the use of an enzyme-tagged antiimmunoglobulin.

A very powerful technique for antibody detection is the Western immunoblot assay. This technique allows the simultaneous detection of antibodies to various antigenic determinants of the microorganism. The antigens from the offending organisms may be purified and well characterized before use in this test, which offers yet another advantage. The antigens are submitted to polyacrylamide gel electrophoresis, which separates them into defined bands. These bands are then transferred to a piece of nitrocellulose paper. This paper is reacted with patient serum. Each antigen against which the patient has antibodies will bind to its respective antigen and be immobilized on the paper. These immobilized antigen-antibody complexes are then detected by using an enzyme-linked anti-human immunoglobulin preparation. The enzyme labeled antigen-antibody complex is detected by reaction with its chromogenic substrate.

The use of immunologic detection and identification techniques to diagnose microbial diseases has a historical precedent that goes back to the early serologic tests for syphilis. In current use, these techniques have the advantages of speed and specificity and do not require that the organisms in specimens be viable. The simplicity of these tests have permitted some to be approved for use in the physician's office, thus improving the ability to provide a patient with a speedy diagnosis.

Genetic Features

The ultimate difference between microorganisms is the specific sequence of nucleotides in the genome of the organism. When techniques become available to identify specific gene sequences, it will be possible to identify microorganisms at the most fundamental level. Currently, restriction endonucleases provide a method for characterizing the genome of microorganisms, although the utility and availability of these techniques for diagnostic purposes need further development. Restriction endonucleases are able to cut DNA at specific nucleotide sequences. The resulting pieces can be separated by electrophoresis into distinct patterns of bands and these bands transferred to a nylon membrane. The ability of polynucleotide sequences to bind complimentary polynucleotide sequences allows for the construction of gene probes. These are specific polynucleotide sequences to which a radioactive label has been attached. When specific gene probes are reacted with the restriction fragments on the nylon membrane, they bind to those gene fragments that have a complimentary nucleotide sequence and in the process label those specific sequences. The radioactivity is detected by exposure of the nylon sheet to X-ray film, which is then developed to show the location of bands.

Presently, these types of gene-probe tests are being developed by numerous biotechnology and pharmaceutical firms. It is not clear how soon the expected revolution in the way we identify microorganisms in clinical samples will occur, but undoubtedly such innovations will become clinically useful in the near future.

SUMMARY

In this chapter the basic biology of microorganisms has been discussed. It should be noted that this has been a very selective review that concentrates on those microbial characteristics that have the greatest relevance to infectious disease problems in clinical obstetrics and gynecology. In view of this fact, more detailed information on many of the subjects discussed can be found in standard microbiology texts.

Just as Lister admonished his students to "see" microbes as enemies, it is important for the physician today to see microbes with modern clarity—not only as the bearers of illness but also as a diverse, not uniformly sinister, menagerie of organisms. Only by holding a balanced view of the microbial world can a rational approach to disease be achieved.

2

The Host in Obstetric and Gynecologic Infections: Infection and Immunity in the Female Genital Tract

HOST-PARASITE INTERACTIONS

In teaching medical students about the various microorganisms that cause infectious diseases, the very important role the host plays in the process is often overlooked or underemphasized. This leads to the problem of oversimplifying the classification of microorganisms into two groups: pathogens and nonpathogens. In this chapter, the balance between host defenses and microbial virulence will be emphasized.

Infectious disease is above all an interaction of microorganisms with the host. It is also an unbalanced state in which the bacteria are attempting unrestricted growth and the host is attempting to eradicate the offender. The result for the microorganisms at the site of infection is that their numbers will be reduced by host factors. The rapid growth of pathogens in the host results in the elaboration of bacterial by-products that may damage the host, or the host's own response to the organism may cause damage to the host as well as to the microorganism. Thus, the final result of the unbalanced interaction between host and microbe is the elicitation of symptoms in the host.

The interaction between host and microorganism is still not fully understood at the molecular level, although many aspects have been studied. Nevertheless the interaction was described by Theobald Smith early in this century. This equation summarizes the theoretic interactions that occur in the process of disease. Smith's equation,

$$\text{disease} = (\text{number of organisms} \times \text{virulence of organisms})/\text{host defenses},$$

contains all of the elements that are involved in the production of disease. This relationship takes into account the possibility that a small number of highly virulent organisms may cause disease and that a large number of low-virulence organisms may cause disease. It also allows for the possibility that a small number of moderately virulent organisms may produce disease in a host with compromised defenses. Above all, the point that must be emphasized is that both host and microorganism contribute to the disease process.

MECHANISMS OF PATHOGENICITY

The pathogenicity of microorganisms is a fascinating concept from the standpoint of the survival of the microorganism. It is easy to understand why a microorganism might associate with a human host. The environment is warm, moist, secure, and homeostatically regulated, with the nutrients supplied by the host. Yet the microorganism cannot be either too aggressive or too passive. If the organism is too passive, it might not be able to establish itself in the host. If the organism is too aggressive, it will kill its host and have no habitat. The most beneficial arrangement, therefore, is for the organism to infect the host in such a way that the host response to the organism is minimal.

This need for balance between passivity and aggression is perhaps best illustrated by the long-term incidence of infectious diseases such as respiratory tuberculosis, diphtheria, measles and pertussis. All of these diseases were declining before such medical innovations as antibiotics, antitoxins, and vaccines or even the ability to culture the offending agent. The decline in the incidence of these diseases before the existence of the medical knowledge needed to mount a rational assault against them may indicate either that the offending organisms were too pathogenic and killed too many of the susceptible hosts, or that the population of susceptible hosts was becoming more resistant as a result of the improvement of living conditions and nutrition.

In summarizing the attributes that confer virulence upon an organism, it is appropriate to remember that a broad range of microbial characteristics are encompassed under the rubric of virulence. For convenience, the following summary of virulence characteristics is presented in somewhat arbitrary categories related to the events that are believed to occur during pathogenesis. These events include the ability of the microbe to survive in transit from host to host, attachment to the host tissue, invasion of the host, adaptation to life within the host, proliferation of the microorganism, and damage to the host resulting in the symptoms of infectious disease.

Communicability

The ability of an organism to survive its journey from one host to another is critical in the ongoing existence of disease-causing microbes. Some organisms must chance exposure to the hostility of the environment, which may include desiccation of the organism, ultraviolet radiation, nutrient deprivation, and extremes of temperature. The dangers of exiting one host and entering another are obviated by those organisms that travel enveloped in secretions or are transmitted by direct contact from host to host. This type of transit is typical of sexually transmitted pathogens, which have a very efficient mechanism of obtaining safe passage to a new host.

Adhesiveness

Unless the organism is directly implanted into the site of infection by some breach of anatomical barriers, the microbe must attach itself to a mucosal or skin surface and gain access to deeper tissues from that attachment site. This attribute of adhesiveness is shared by pathogens and normal flora organisms alike. Not surprisingly, many of the normal flora organisms are able to cause infectious disease after obtaining entrance through some break in the mucosal barriers. The role of adhesiveness among both normal flora organisms and mucosal pathogens has been studied extensively. Although this role cannot be reviewed in detail here, it deserves brief comment.

Adhesive phenomena among organisms that cause sexually transmitted disease are exemplified by work on *Neisseria gonorrhoeae,* which has been shown to display histotropism for columnar epithelial cells. The attachment of gonococci is associated with the presence of pili on the bacterial surface, which has been considered to be a sine qua non of pathogenicity. Adhesion to epithelial surfaces, however, may also involve outer membrane protein II, which has also been associated with virulence. The possibility of producing a vaccine to prevent adhesion, based on antibody formation against pilus proteins, has been one of the goals of microbiologists working on the gonorrhea problem. This goal emphasizes the significance attributed to attachment in the overall disease process.

The mechanisms of adhesion are not identical for all microorganisms that attach to epithelia. Some organisms, such as *Escherichia coli,* attach by means of distinct fimbrial structures that adhere to mannose residues on uroepithelial cells. Other organisms, such as streptococci, do not attach by means of appendages; rather, the surface lipoteichoic acid interacts with fibronectin on epithelial cells. The syphilis spirochete also attaches to fibronectin. Many bacteria are known to attach to the glycocalyx of mammalian epithelial cells, but the precise bacterial adhesin molecules have not been characterized. *Gardnerella vaginalis* is well known for its attachment to vaginal epithelial cells (clue cells), but the adhesin and receptor have not been identified. The attachment of *Mobiluncus* to vaginal epithelium has been demonstrated by electron microscopy and resembles the attachment of normal flora organisms to epithelial surfaces. Some of the putative attachment proteins of *Candida albicans* have been isolated and appear to be more prominent in germinated than in yeast phase cells. Other organisms that possess adhesive factors will undoubtedly be identified in the future. These factors may be as important in maintaining the stability of the normal flora as they are in determining the virulence of pathogenic organisms. Perhaps in the larger context they represent variations on a single theme: a microorganism developing a relationship with the host that is suitable for the survival of both partners.

Invasiveness

The next step in the genesis of infectious disease is invasion of tissues by the microorganisms. There are many methods whereby microorganisms enter the host tissues. For example, the gonococcus is engulfed by nonprofessional phagocytes as it moves from the epithelial surface toward the deeper tissue layers. Other microorganisms simply reach distant tissue sites by destruction of tissue structures. Organisms that produce proteases, collagenases, or hyaluronidase are capable of disorganizing the integrity of cellular connections, allowing the spread of an infectious process. Another interesting hypothesis has been proposed regarding microbial spread in the female genital tract, namely, that certain organisms have the ability to attach to spermatozoa, which carry the bacteria to the upper genital tract. The attachment of organisms to sperm has been documented, but the importance of this phenomenon in actual upper-tract infections has not been established.

Adaptability

Once in the host, whether at an epithelial surface or in deeper tissue layers, the microorganism must be able to adapt to the microenvironment. Little is known about what specific adaptations an organism must make as it attempts to grow rapidly in the host's tissues, although animal studies have demonstrated that adaptation does occur. For example, the lethal dose of a particular species of microorganism that has been subcultured for many generations on artificial media may be greater than the lethal dose of the same species of microbe freshly isolated from an infected animal. The process of attenuation of virulence has in fact been exploited to produce living vaccine strains of microorganisms, as is the case with the vaccine strain of *Mycobacterium tuberculosis*.

The essence of adaptation to the environment within the host is evasion of the host's antimicrobial defenses. A microorganism must avoid the bacteriolytic properties of serum, evade uptake or destruction by phagocytic cells, and successfully compete for essential nutrients such as iron. The production of capsular material will mask surface antigens and provide a physical impediment to phagocytosis. If a microorganism undergoes a shift in the major antigens it displays on its surface, the host will be caught making the wrong antibodies and the microorganism will avoid destruction. Organisms that display antigens that have a high degree of cross-reactivity with host antigens may either go unrecognized or may cause the host to damage itself rather than the microorganism. Staphylococci possess a protein (protein A) that binds immunoglobulin in such a way as to render it ineffective, whereas other mucosal pathogens produce proteases that degrade immunoglobulin A (IgA), the dominant antibody of secretions. Some organisms also produce surface substances that render the complement-mediated system of bacteriolysis ineffective. The ability to survive and multiply within phagocytic cells represents an additional strategy of adaptation to the environment within the host.

This brief summary of microbial adaptations to the conditions imposed by the host is not intended to be exhaustive. Rather, it illustrates the diversity of survival strategies used by microorganisms in dealing with the adversities of life as a pathogen.

Proliferation

The ability to undergo rapid growth has many advantages for microorganisms. Such small organisms have little capacity for amassing large stores of nutrients for hard times, and so must use nutrients when they can. Using food to produce large numbers of organisms enhances the chances of the survival of at least some of their numbers during lean times. Rapid proliferation has another advantage as well. The diversity that is generated mainly during replicative cycles allows for the adaptive processes mentioned above to occur quickly. As the environment provides selective pressures, new generations with different phenotypic characteristics arise. The microbial population does not become extinct, as might be the case with much larger organisms placed in a rapidly changing environment.

Rapid proliferation also allows the organism to maintain an advantage over the host's defenses. Phagocytosis will reduce the number of viable organisms; therefore, it is important for the organism to maintain its numbers through replication. There are certain soluble host defense factors that interact with bacterial cells according to the laws of mass action. Thus, the more bacterial cells there are, the fewer molecular defense molecules there are to act on each bacterium. Clearly, it is advantageous for the microorganism to have the capacity for rapid growth.

Toxigenicity

The presence of microorganisms within a host or even within the host's tissues is not synonymous with disease. Disease results from symptoms, which are due to either the host's reaction to the microorganisms or the toxic products of the microorganisms. At the mention of bacterial toxins, classical examples such as the endotoxin or lipopolysaccharide produced by gram-negative bacteria or the tetanus exotoxin come too quickly to mind. However, it should be realized that some of the toxic properties of microorganisms are not the result of compounds that would be ordinarily classified as toxins. For example, succinic acid, which is a by-product of the growth of members of the anaerobic genus *Bacteroides*, is able to inhibit the functional capability of phagocytes. This discovery demonstrates that bacteria can produce adverse effects on the host without the highly toxic substances that are found in some of the most virulent organisms.

It is beyond the scope of this chapter to catalog all of the known bacterial toxins. Such a catalog would not be germane to this work because many of the organisms that produce the most potent toxins are irrelevant to obstetric and gynecologic infections. Specific toxins will be mentioned in relation to specific organisms in later chapters, but at this point some of the ways in which bacterial toxins interact with the host's defense capabilities will be summarized.

A variety of bacterial exotoxins are able to disrupt or otherwise interfere with the integrity of cell membranes. Such cytolytic toxins may aid the growth of the bacterium by causing the release of the intracellular contents of host cells, including erythrocytes, which are a rich source of the iron needed by bacteria for growth and virulence. The resulting tissue destruction can cause symptoms. Moreover, cytolytic toxins may cause the lysis of phagocytic cells, which play such a critical role in the control of the infectious process.

Spreading factors such as proteases and sialidases can also be categorized as local toxic factors. They may aid the organism in its spread into areas where nutrients may not be depleted, pH may be more favorable, or where growth-retarding bacterial metabolites are not present in high concentration.

Bacterial endotoxin, which is the lipopolysaccharide associated with the outer membrane of gram-negative bacteria, is a very complicated compound both chemically and physiologically. It does possess intrinsic toxic activity and in sufficiently high concentrations will affect the host's physiology in ways that produce the condition of septic shock. Endotoxin may also act synergistically with other toxins to enhance their biologic effects. Ironically, endotoxin may produce signals that activate the immune system, sometimes to the detriment of the host. Fever caused by the release of interleukin-1 and tumor necrosis factor from white cells is one of the signs of exposure to endotoxin.

A substantial number of soluble mediators are known to be involved in the acute-phase response to injury, including infection, trauma, and exposure to endotoxin. The acute phase is characterized by fever, synthesis of C-reactive protein by liver, and altered concentrations of serum copper, zinc, and iron. The responsible mediators include interleukin-1α and interleukin-1β, interleukin-6, tumor necrosis factors, and interferon-γ. These mediators appear not only to stimulate the immune response, but also to link the neuroendocrine system to the immune system.

Endotoxin has immunostimulant properties, as well as the ability to activate the alternate pathway of complement. Endotoxin also can act as a B-lymphocyte mitogen and can activate macrophages, resulting in increased production of lysosomal enzymes and lymphokines. Many of these activities in the locale of the infection cause enhanced inflammation.

Thus far, the discussion of the toxic properties of microorganisms and the symptoms they cause has focused on local effects of the toxins. Some toxins act at locations remote from the site of bacterial growth. The advantage to the microorganism of producing distantly acting toxins is difficult to explain. As is the case with local toxins, some of the symptoms may be due to the direct effect of the toxin and some of the symptoms may be due to the physiologic response of the host to the toxin. The hemodynamic changes that are seen in cases of endotoxic shock exemplify the systemic changes in physiology related to lipopolysaccharide.

ORGANIZATION OF THE HOST'S ANTIMICROBIAL DEFENSE

Because the defenses of the normal host are as important a determinant of infectious disease as are the properties of microorganisms, it is appropriate to review the mechanisms that provide the host with resistance to disease. It will then be possible to look at these mechanisms in relation to the specific anatomic structures of the female pelvis. The major categories of host defense that will be considered include anatomical barriers, nonspecific cellular and humoral mechanisms of defense, immunologically specific mechanisms including cellular and humoral aspects, and local immunity.

Anatomic Barriers

Some microorganisms, such as those that cause vaginal infections, produce symptoms on epithelial surfaces. Others require access to deeper tissues to produce symptoms. For those bacteria that require access to deeper tissues to produce symptoms of infectious disease, the primary impediment to their entrance into the normally sterile tissues of the host is the intact skin or mucosal epithelia. An important anatomic tissue barrier well known to the obstetrician is the chorioamniotic membrane. The importance of this structure in preventing intraamniotic sepsis is well illustrated by cases of fulminant sepsis, which may occur after amniorrhexis.

The microorganisms that make up the normal flora of the gut or vagina and endocervix include many species that are able to produce disease in the subepithelial tissues after the integrity of the tissue is compromised by a surgical procedure. However, these organisms are benign when confined to the tissues they normally colonize and furthermore do not possess the ability to make their way into the deeper tissues.

Usually barrier functions are performed by intact tissues. But it is possible that barrier functions may be performed in other ways. As will be described in greater detail in chapter 3, the normal flora itself may function as a barrier, preventing colonization with more virulent microorganisms. The mucus sheet on the respiratory epithelium and the mucus that fills the endocervical canal may also function as barriers.

Nonspecific Cellular Defenses

When microorganisms gain access to the tissues or the bloodstream of the host, they are met by phagocytic cells that are able to ingest and destroy the bacteria. Two classes of so-called "professional" phagocytic cells make up the host's nonspecific cellular defense. The polymorphonuclear neutrophilic leukocytes are about evenly divided between circulation in the bloodstream and attachment to the vascular endothelium. In the event of tissue injury, these cells leave the vascular sites and move in as the first wave of cellular defenses. If the inflammatory process becomes chronic, however, the macrophages enter the injured site. One role of the macrophages is to remove blood-borne microorganisms from the circulation. The macrophages are affixed to the liver and spleen and form the reticuloendothelial system. Many tissues of the body contain macrophages. These are termed "wandering macrophages" and are not particularly aggressive in the early inflammatory reaction, although their production of the lymphokines interleukin-1 and tumor necrosis factor causes recruitment of additional activated polymorphonuclear leukocytes to the injured tissue. Macrophages may be activated to a more aggressive state by the lymphokines.

The activation of macrophages by the lymphokines, which are secreted by antigen-specific lymphocytes, confers an immunologic specificity on the otherwise nonspecific macrophage defense system. The nonspecific neutrophilic phagocytes do possess a specific aspect in that phagocytosis of particles proceeds more readily when the bacterium has been opsonized. This is a process by which specific antibodies bind to the bacterial cell, allowing the phagocyte a better chance to attach to its surface. Thus, it should be recognized that although defenses mounted by polymorphonuclear and mononuclear phagocytes may be nonspecific, they are more efficient when coupled with the host's immunologically specific recognition. This fact emphasizes the unity of the various elements of the host defense system.

As understanding of phagocytosis has developed, it has become increasingly apparent that the phagocytes engage in a very complex process that includes more than swallowing and digesting particles. The process is preceded by chemotaxis, which is the attraction of phagocytes to the site of infection by chemorecognition of, and directed movement of the phagocytes toward, the microorganism.

One of the earliest events eliciting the phagocytic defense is the inflammatory reaction, which results in the recruitment of phagocytes to the site of injury. In contrast to mammalian cells, bacterial protein synthesis begins with the unusual amino acid *N*-

formylmethionine, followed by the other amino acids that make up the whole of the protein. The final step in the maturation of the nascent peptide is the cleavage of the *N*-formylmethionine. Interestingly, small peptides that have *N*-formylmethionine as their N terminus are strongly chemotactic for white cells. The host, therefore, possesses a system for detecting the presence of bacteria independent of specific antigenic recognition.

White blood cells respond chemotactically to complement cleavage products and to other signals, including leukotrienes, which are elaborated from other phagocytes at the site of inflammation. A fuller description of complement-mediated host defenses will be included in the discussion of nonspecific humoral defenses.

After being attracted to the microorganism, the phagocytic cell attaches to the surface of the microorganism, extends pseudopodia around it, and engulfs the microbe in a membrane-bound vesicle called the phagosome. Ligand molecules called opsonins coat the surface of the microorganism, allowing the phagocytic cell to attach; this process is called opsonization. Specific antibody is one of the best known of the opsonic factors. As the variable region of opsonic antibody binds to the microorganism, the Fc region of the immunoglobulin is free to interact with Fc receptors on the phagocytic cell membrane. The progressive attachment of more phagocyte receptors to the opsonic molecules allows the envelopment of the microorganism in a process that resembles the action of a zipper.

The process of engulfment is accompanied by triggering intracellular metabolic activities that are activated during particle ingestion. The metabolic activities generate toxic substances that kill the ingested microorganisms. These substances are referred to as the oxygen-dependent killing mechanisms. Specifically, the oxidation of glucose through the hexose monophosphate shunt generates oxygen-containing free radicals that may be directly toxic to microorganisms. In addition, the enzyme myeloperoxidase in the presence of chloride and hydrogen peroxide (generated during oxidative metabolism) causes peroxidation of membrane lipids of the bacterial cells or the generation of toxic hypochlorous anions, resulting in death of the bacteria.

The bacteria contained in intracellular phagosomes are exposed to a constellation of hydrolytic enzymes when lysosomes coalesce with phagosomes to form the phagolysosomes. Two types of leukocyte lysosomes exist. One is the specific granule and contains oxidases, lactoferrin, and lysozyme. The azurophilic granules contain a complex mixture of antibacterial substances that include antibacterial cationic peptides and compounds known as defensins, as well as myeloperoxidase. The fusion of lysosomes with phagosomes brings the various antimicrobial properties of the phagocyte into proximity with the microorganism, making the process of intracellular killing particularly efficient.

The foregoing discussion shows that each step of the phagocytic process involves extracellular and intracellular recognition, signalling, and control and that the process of disposing of microorganisms is exceedingly complex. Because of the variety of steps involved, there are many potential ways in which the phagocytic defenses can be stimulated or suppressed. These issues become important when the genesis of infectious diseases in surgical or oncology patients is considered.

Nonspecific Humoral Defenses

All of the soluble factors that are involved in host defense are synthesized by various cells or tissues of the body. However, inasmuch as their role in defense against infection is as soluble factors, they are for convenience described as humoral factors. Some of these factors are elaborated by the phagocytic cells and are active both in the phagolysosomes and in serum, interstitial fluids, or in the interstitial fluids at the site of an inflammatory reaction.

Antibacterial activity may be derived from the ability of the host to withhold iron from bacteria. Transferrin normally serves the purpose of iron transport in serum and leaves virtually no iron free to support bacterial growth. In fact, the degree of iron saturation of transferrin (about 30-40%) gives serum transferrin considerable reserve capacity for acquiring any excess iron that may enter the bloodstream. Lactoferrin, like transferrin, is an iron-binding protein. It was originally discovered in milk (hence its name) but is found in leukocyte granules and in various secretions. In addition to its iron-withholding ability, this compound may be directly toxic to some microorganisms when it is associated with iron.

Lysozyme occurs in secretions and has the important property of hydrolyzing the peptidoglycan material of the bacterial cell wall. Generally, sufficient concentrations of this enzyme are not present in body secretions to cause bacterial lysis. However, in the presence of other antimicrobial substances, lysozyme may contribute to destruction of microorganisms as the result of synergy among various factors.

Complement is a complex system of proteins in serum, some of which are associated with cell membranes. The main function of this system with respect to host defense is to cause the lysis and death of bacterial cells through membrane damage. Because membrane damage is one of the end results of complement action, virus-infected cells that display foreign antigens on their surfaces and enveloped viruses, which are surrounded by membrane from the host in which they replicate, can be damaged by the membrane effects of complement. Complement, however, has many other actions, some of which also are part of the host's defense. Some complement components are chemotactic, and other components possess an opsonic function. The complement system also is able to enhance inflammatory reactions by the release of interleukin-1, tumor necrosis factor, and anaphylatoxin.

The various actions of complement are derived from the diverse components of this system or from their reaction products. The actions of complement can be generated through one of two pathways that later converge to produce a membrane attack complex that is ultimately responsible for lysis of bacterial or other cell membranes.

The classical pathway for complement activation results from the reaction of antibody with antigen. This complex allows some of the early complement components to come together in an active form, which initiates a cascade of subsequent reactions. The cascade of reactions includes the generation of opsonic and chemotactic factors before the ultimate formation of the membrane attack complex, thus providing a dual mechanism of protection of the host: stimulation of the phagocytic defense and membrane lysis.

The alternate pathway is indirectly activated by the presence of bacterial surface components. Thus, complement may be activated in a manner that does not involve immunologic specificity. In this case, an endogenous protease cleaves the third component of complement to its active form, which in the absence of bacteria is immediately inactivated by a particular endogenous inhibitor. But in the presence of bacteria, the activated third component is protected from subsequent inactivation by binding to the bacterial cell. The remaining components of complement may then react in sequence, ultimately resulting in bacterial membrane damage.

The complement system has general relevance to bacterial and other infections, but for the gynecologist at least one striking example of its importance is the finding that a congenital deficiency in terminal complement reactants results in a particular susceptibility to gonococcal infection.

Macrophages elaborate a number of modulators of the immune response and other biologic systems. Some of these products may not have a direct antimicrobial effect, although they may contribute indirectly to the overall host defense. The interleukins stimulate various classes of lymphoid cells, which may enhance specific immunologic reactions or cause more rapid production of lymphoid cells from stem cells in the bone marrow. Interleukin-1 is notable because it is involved in the production of the febrile response to infection, which probably enhances host defense in a variety of nonspecific ways. The interferons are also products of various lymphoid cells produced in response to viral infection or exposure to mitogenic substances such as bacterial endotoxin. One of the important activities of interferon is the suppression of viral replication.

Specific Immunity

In addition to the innate nonspecific defenses outlined above, the intact host possesses a multifaceted immune system that specifically recognizes foreign substances and reacts to them. Because this discussion centers on host defense, the specific immunologic reactions, including tolerance and hypersensitivity, will be bypassed and attention will be focused on mechanisms for the immunologic destruction of microorganisms. The ability to recognize foreign substances depends on the functional properties of lymphoid cells. The ultimate effector mechanisms may be either cellular or humoral, although several echelons of cellular functions lead to both the humoral and cell-mediated immune reactions.

In very general terms, the immune response consists of several common events, which are summarized in Fig. 2-1. The host first brings the foreign substance into contact with cells of the immune system. This function is usually reserved for the macrophages, which ingest foreign particles without the benefit of immunologic recognition. As noted previously, one of the mechanisms that macrophages use to identify bacteria is attraction to the *N*-formylated peptides that are characteristic of bacterial protein synthesis. Bacterial lipopolysaccharide or other unique microbial surface substances likewise attract macrophages. When the intracellular mechanisms digest the particle, the immediate goal of microbial destruction is achieved. However, a long-term result also is attained, representing the second phase of the immunologic response.

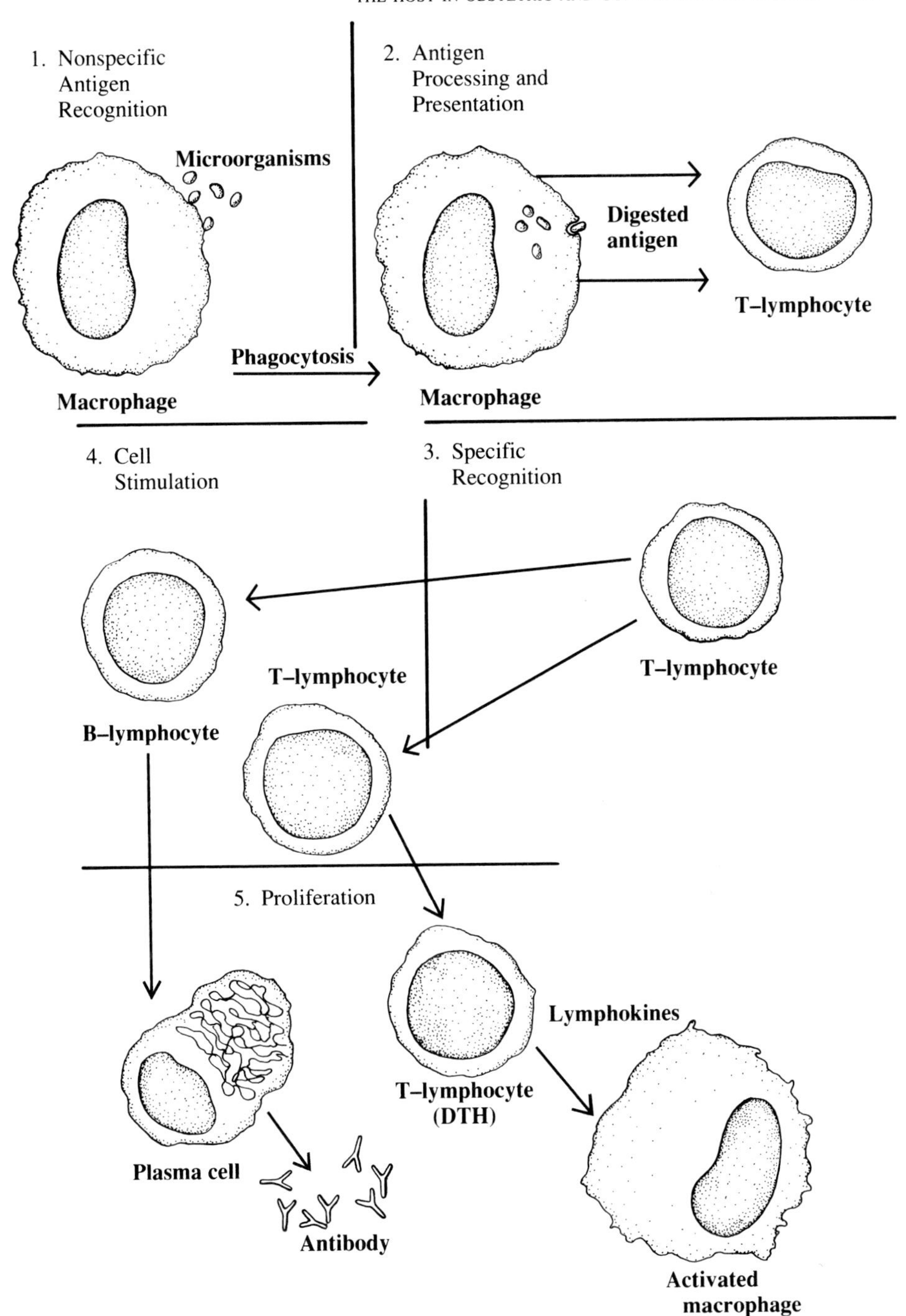

Fig. 2-1. *Elements of the immune response.*

The foreign antigens that are now contained in digested form within the macrophage are transported to the surface of the cell. These antigens, complexed with the type II major histocompatibility complex molecules, are the material that specific T lymphocytes recognize as foreign. This process is described as antigen presentation. The T lymphocytes display a division of labor. Some cells augment or suppress the immune response (T-helper and T-suppressor cells) and some engage in cytotoxic functions. The T-helper cells are usually involved in the recognition of the microbial antigens that are displayed on microbial surfaces along with the class II major histocompatibility complex, and they signal specific B or T lymphocytes to proliferate. A response is produced that is essentially directed at eliminating the foreign antigen, at the same time expanding the population of lymphoid cells specifically capable of recognizing that particular antigen.

The B lymphocytes are immunoglobulin-producing cells and have on their surfaces antigen-specific immunoglobulin molecules. They may recognize antigens directly, but more commonly they rely on T-helper cells to provide a signal that causes them to become antibody-secreting cells. The actual mass production of antibodies results from the differentiation and proliferation of stimulated B lymphocytes into plasma cells that produce antibodies. Because the antibodies are soluble factors, this part of the immune response is described as humoral immunity.

The reader is undoubtedly familiar with the role of antibody in host defense, but a few major points will be stressed. Antibody, which is a functional designation, or immunoglobulin, which is a biochemical descriptor, consists of complex aggregations of polypeptide chains that are responsible for class distinctions among the immunoglobulins. Each type of immunoglobulin consists of heavy and light chains. Immunoglobulins are classified according to the nature of the heavy chains (IgG, IgM, IgA, IgD, and IgE). For a complete discussion of the structural and functional differences among these classes, the reader should refer to a standard immunology text.

Each immunoglobulin contains regions with relatively constant primary amino acid sequences. These regions may attach to certain cellular receptors such as the Fc receptor present on various types of white blood cells. Each immunoglobulin also contains a region of variable structure that provides for an antigen-binding site tailored to combine very specifically with any of a number of antigens. It should be noted that the immunoglobulins produced by a specific B-cell clone will be identical in terms of the specificity of antibody secreted, although the body as a whole has a repertoire of lymphocytes that can respond immunologically to perhaps millions of antigenic determinants.

The cell-mediated part of the immune system involves cellular effectors. Several types of cell-mediated immune reactions can be catalogued, including rejection of tissue grafts, destruction of tumors, and antimicrobial immunity. The latter is of primary importance for the purpose of this chapter, although cytotoxic destruction of virus-infected cells also falls within the scope of the microbiology of obstetric and gynecologic infections.

Cell-mediated immunity involves the recognition of antigen by T lymphocytes. Upon stimulation by antigen, these cells undergo rapid proliferation analogous to that characteristic of the blastogenesis described for B cells. This proliferation results in an increased number of effector cell clones (the delayed-type hypersensitivity, or T-dth, cells) and enhanced production of lymphokines and monokines. Of primary interest from the standpoint of antimicrobial effect is the generation of monokines. These substances have among their functions the stimulation of macrophages. The macrophages so affected are described as activated or "angry" and possess heightened antimicrobial activity. Cytotoxic T cells (T-c cells) are generated as a result of antigen stimulation, and they kill target cells such as tumor cells by contact with the target and release of lysins that destroy the target cell.

For the sake of completeness, it should be noted that cell-mediated immunity may also involve cells belonging to a class known as the large granular lymphocytes, which include the natural killer cells and the killer cells. The former appear to be stimulated by interferon and are able to kill tumor cells. They do not have surface markers characteristic of T or B lymphocytes. The killer cell is also considered to be a non-B, non-T cell type and is involved in antibody (IgG)-dependent cell-mediated cytotoxicity presumably due in part to the presence of an Fc receptor on the surface of the cell.

Local Specific Immunity

With this category of protective mechanisms, our discussion of host defense comes full circle. This discussion began with the role of anatomic barriers, and here it will be shown that the effectiveness of the anatomic barriers is improved by the ability of the host to produce specific antibodies to protect those tissues. The major antibody class found in secretions is secretory IgA, which differs from the circulating form of IgA. The basic IgA monomer consists of two light chains (κ or λ chains, as found in IgG) and two heavy (α) chains. The secretory IgA, which is found on mucosal surfaces or in body secretions (tears, lung, gastrointestinal lumen, uterine cervix, colostrum, and milk), is composed of two IgA monomers linked by a peptide designated as the J chain. Also bridging the monomers is an accessory peptide identified as the secretory component. Differences in α chains give rise to the subclasses designated as IgA1 and IgA2, which are present in secretions in approximately equal abundance. The former is more susceptible to proteolytic cleavage by IgA proteases elaborated by some mucosal pathogens.

The function of IgA in the secretions is believed to relate primarily to blocking the attachment of pathogenic organisms to mucosal epithelium. There may be other mechanisms whereby IgA defends mucosa, such as synergistic interaction with lysozyme and transferrin-mediated antibacterial effects. One of the important characteristics of the local immune system is that stimulation of lymphoid cells that elaborate IgA antibody occurs at sites very near to the source of the antigen. The lymphoid tissues that underlie the mucosa are designated by the acronym MALT (mucosa-associated lymphoid tissue). These tissues are stimulated by local antigen and in turn secrete local antibody. The host response to the stimulus is therefore characterized by antigenic specificity and site specificity.

HOST DEFENSES OF THE FEMALE

Systemic Defenses

Before discussing specific anatomic structures unique to the female patient, it should be briefly mentioned that the entire immune system appears to benefit from being in a female host. Female babies are intrinsically more resistant to infection than male babies. Thus, the argument that the shorter life expectancy of males compared with females is due to hormone factors and stressful or hazardous occupations held by men is refuted. A locus on the X chromosome that is involved in IgM synthesis may account for some of the differences. Compared to males, females also have intrinsically higher levels of circulating immunoglobulins. Interestingly, many of the currently identified immune deficiency syndromes are X-linked (for example Bruton's agammaglobulinemia) and appear in boys. Although males are more commonly afflicted with immune deficiencies, females are more likely to suffer from diseases related to exuberant immunity (eg, myasthenia gravis, rheumatoid arthritis, and Hashimoto thyroiditis).

Although the sex steroids are not responsible for differences in early susceptibility to infection, they may play a role later. Again, females have the advantage. The effectiveness of the phagocytic response of reticuloendothelial cells is known to be augmented by estrogens, according to studies with explanted animal liver tissue. During pregnancy, peripheral monocytes show an increased ability to ingest latex particles under experimental conditions.

As suggested by studies on phagocytic function, some of the physiologic adaptations to pregnancy may alter certain aspects of systemic host defense. For example, during pregnancy the ß-globulin fraction of the serum increases and with it the concentration of transferrin increases, whereas the iron saturation of transferrin diminishes. This is consistent with an increased protective capacity of the serum. The endogenous levels of complement factors increase during pregnancy, and the absolute number of circulating phagocytic cells in the peripheral blood also increases.

Together, the host defenses of the female display some advantages over those of the male, advantages that pregnancy only serves to enhance. This fact makes teleological sense from the standpoint of the mother providing a safe environment for the fetus, not only in terms of nutritional support but also in terms of protection from the hazards of the pathogenic microbial world. The primary area in which host defense is diminished during pregnancy appears to be cell-mediated immunity, which can be adequately explained by the need of the mother (who is partially antigenically different from her fetus) to avoid immunologic rejection of the conceptus.

Vulvar and Vaginal Epithelium

The intact epithelia of the vulva and vagina provide a barrier to the invasion of microorganisms, although some pathogens have the ability to invade these tissues. One of the main reasons why the human skin has a limited flora is its dryness. The tissue of the vulva naturally has a higher moisture content, which makes it more susceptible to colonization by certain organisms and probably to intertriginous infection by yeasts. When ulcerations occur, as in the case of herpetic infection, the barrier function of the vulvar epithelium is compromised and a secondary bacterial infection may ensue.

The barrier function of the vaginal epithelium is obvious in view of the abundant flora that colonizes it without invading the deeper tissue layers. Indeed, the vaginal flora itself, which will be discussed in detail in chapter 3, is considered to be part of the protective mechanism of the vaginal epithelium. The vaginal lumen includes adequate moisture to support an abundant flora on a stratified epithelium. Typically the pH at this site is low. There is disagreement as to whether this low pH is the result of bacterial metabolism or whether the pH is naturally low when the vagina is sufficiently estrogenized. The latter seems likely in view of the observation that infants who have received estrogen transplacentally but who have not yet been colonized by acid-producing bacteria have low vaginal pH. The vagina is also characterized by a low oxidation-reduction potential. It is believed that additional oxygen can enter the vaginal environment with the insertion of a tampon.

The tissues underlying the vaginal epithelium provide blood supply to the tissue and provide a source of leukocytes in the vaginal lumen. Normally, the thickness of the superficial layers restricts the abundance of white cells on the vaginal lumenal surface. In some cases of vaginitis, leukocytosis can be extensive and indicates the availability of an inducible host defense.

The soluble host defense factors present in the fluid that bathes the vaginal epithelium have not been established. The vagina is not a secretory organ; therefore, the secretory immune protection must be derived from the secretions of the endocervix. Dilution of these cervical defense factors may occur as the secretion is distributed across the considerable surface area of the vagina and mixes with fluids derived by transudation from the underlying vasculature.

Uterine Cervix

The endocervical canal is the portal of entry for microorganisms from the external environment. The canal is contiguous with the vagina and the internal environment of the uterus, which ultimately communicates with the peritoneal cavity. The endocervix is lined with glands and normally contains an accumulation of mucus. It is often claimed that this mucus forms a physical barrier between the lower and upper genital tract, although it is surely an evanescent barrier. Certainly during menstruation the barrier function is diminished, and during pregnancy and in the puerperium the dilatation of the cervix may preclude the existence of a stable physical mucus barrier. It is probably more appropriate to think of the mucus as a biochemical impediment rather than a physical barrier. The endocervix contains secretory immunoglobulins and very high levels of lysozyme. The lysozyme content may be 100-fold greater than that found in serum. Because of the limited amounts of cervical mucus, the present understanding of how it interacts with specific microorganisms is limited to speculation. However, because the uterus apparently is normally sterile and the vagina has an abundant flora, a biochemical or physical-barrier function of the uterine cervix is implied.

Nongravid Uterus

When salpingitis occurs, it is usually the result of organisms, initially acquired in the lower tract, ascending into the uterus and reaching the fallopian tubes by contiguous spread. This occurrence coupled, with the knowledge that the uterus is not ordinarily contaminated with bacteria, raises the question of whether the uterine lumen has an intact host defense system.

Except in the unlikely event that the uterus becomes infected from a blood-borne source, it is the endometrium that interacts with infectious agents. The endometrium consists of the basalis and functionalis layers, the latter being profoundly affected by the changes in hormones throughout the menstrual cycle. After menstruation, most of the functionalis layer is shed and regrowth begins almost immediately, even before menses cease. The regenerated layer consists of cuboidal epithelial cells with little inflammatory reaction; the macrophages and neutrophils are mainly confined to the stromal tissue rather than being exposed on the lumenal surface. There is no reason to believe that these phagocytic cells would not be chemotactically attracted to the lumen if challenged with a microbial stimulus.

During most of the menstrual cycle, the endometrium is well epithelialized, which probably provides a sound barrier against infection. At menstruation, blood and serous fluids provide additional nutrients, and the denuded surface may be better suited for microbial proliferation than at other times during the cycle. This may account for the temporal association of pelvic inflammatory disease with the menses.

Although infection is more often associated with the endometrium, the myometrium can become infected as a sequel to surgical procedures or as a consequence of postpartum or postabortion endometritis. Adequate perfusion of the tissue and lymphatic drainage are probably the major protective attributes of the myometrium. The lymphatic drainage is mainly associated with the myometrium and extends only a small distance into the basalis layer of the endometrium.

Fetus

While in utero, the fetus depends in large measure on the maternal host defenses for protection. The fetal blood supply is isolated from the maternal blood supply by the placenta, which is a substantial but imperfect barrier. As later chapters will describe, only a limited number of blood-borne pathogens have the ability to cross the placenta to reach the fetal bloodstream. Both bacterial and viral agents are able to act as transplacental pathogens, even though viral agents are incriminated more often. The protection of the fetus from hematogenously acquired infections therefore depends in the first instance on the adequacy of the maternal defenses.

A second route to fetal infection is by upward spread of microorganisms from the mother's lower genital tract to the gravid uterus. In such cases, the barrier function of intact fetal membranes appears to be among the most significant protective mechanisms. Historically, the genesis of fulminant chorioamniotic infection, intraamniotic sepsis, and fetal involvement or death have resulted from the rupture of the fetal membranes before the onset of labor. However, in view of the fact that conservative noninvasive management of the patient with prematurely ruptured membranes frequently results in good outcomes for both the mother and her infant, it seems likely that other host defense factors may also be involved. The obverse of this situation is seen among women in Third World countries or among the urban, ethnic poor in the United States, who probably possess less adequate constitutional defenses and who are very prone to fulminant infection after rupture of fetal membranes. Thus, the intact membranes probably represent part of the natural defense of the fetus.

When microorganisms gain access to the amniotic fluid, they may encounter humoral factors that prevent unbridled proliferation of bacterial growth. Lysozyme, low levels of immunoglobulin, transferrin, complement components, anonymous antibacterial substances, and white blood cells all may be present in the amniotic fluid and may influence bacterial or viral survival. Quantitative bacterial cultures in women who have experienced amniorrhexis before labor suggest that if the bacterial contamination of the amniotic fluid amounts to more than approximately 1,000 viable bacteria per milliliter, the sequelae of infection are likely to occur in the mother or the infant. It appears from these observations that the protective capacity of the amniotic fluid may be overwhelmed by large numbers of bacteria, but minimal contamination may be effectively controlled. The specific organisms may be of importance as well, since the virulence of various bacterial species that may gain access to the amniotic fluid differs among bacterial strains.

Fallopian Tubes and Ovaries

The adnexa have not really been evaluated with regard to specific host defenses, but they probably contain the major host defenses characteristic of a vascularized tissue. Tuboovarian abscess (also referred to as tuboovarian complex, although this is not an appropriately descriptive term) is the main infectious problem associated with this structure. The fact that an abscess can form and persist for an extended period of time indicates that host defenses are operative. It is not clear what role the process of ovulation plays in the susceptibility of the ovary to infection, but as the follicle ruptures a small amount of bleeding occurs, which may provide a site where infection could arise.

The inflammatory response in the fallopian tubes is problematic. It is, of course, the response that is directed at eliminating the infection. Yet as a result of the intense inflammation, as well as the cytotoxic effects of the microorganisms involved in the tubal infection, damage is done to the tissue. This damage may result in nonpatency of the tubes with attendant infertility. The fallopian tubes are important structures because the fimbriated end communicates with the intraabdominal environment. Blood, pus, and microorganisms present in the tubes may escape into the abdomen by this route. It may be speculated that if it were not for host defense factors present in the tubes, tuboovarian abscess and peritonitis would be much more common than is presently observed.

SUMMARY

In this chapter, emphasis has been placed on the concept that the interplay between the virulence attributes of microorganisms and the host resistance attributes of the female are responsible for the genesis of disease. The old concept that microbes can be categorized as pathogens and nonpathogens is passé, inasmuch as almost any organism may cause disease in a host with diminished resistance or in a normally sterile site if implanted in sufficient numbers. Those organisms that have the reputation of being pathogens may cause disease fairly consistently because they possess potent virulence attributes. The host has been shown to maintain an impressive array of defensive strategies to deal with microbial invasion. These defenses are present not only as systemic mechanisms of protection, but also are active in the tissues and structures of the female pelvis. Although the reaction of the host to primary invasion by microorganisms involves constitutive defenses, specific inducible immunity also comes into play to complete the defensive attributes of the host.

3

Microbial Composition of the Lower Female Genital Tract

SIGNIFICANCE OF THE GENITAL MICROFLORA

One of the constant pursuits of the scientist is to find unifying concepts that explain the phenomena of nature. For the student of infectious diseases that primarily affect the pelvic structures of women, the composition and behavior of the vaginal flora become the central unifying features of these infectious processes.

Two general categories of infectious diseases involve the normal vaginal flora. First, the normal flora contains organisms that contaminate surgical wounds or may contaminate the endometrium after parturition. Second, the vaginal flora may contain exogenous microorganisms, such as those acquired by sexual intercourse. These newcomers must compete with the established flora to survive and produce disease. Likewise, vaginitis either arises from an alteration of the normal flora or is superimposed on the normal flora. A summary of the types of infection associated with the normal flora for anaerobic species is shown in Fig. 3-1.

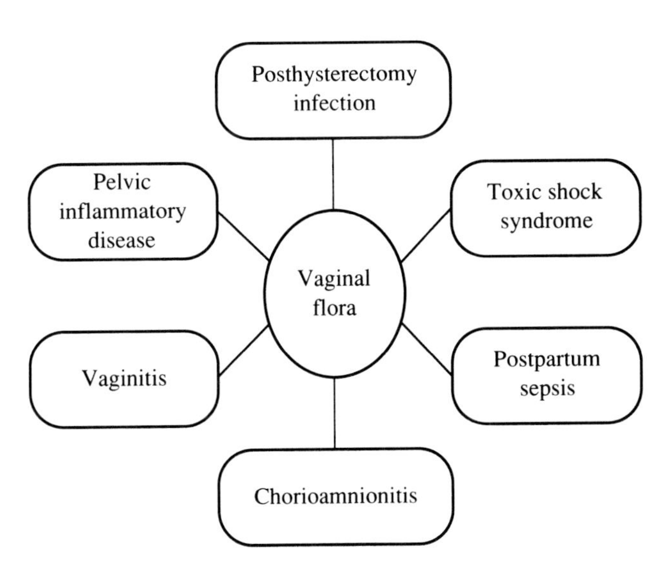

Fig. 3-1. *Infections that can be caused by the normal vaginal flora.*

PRINCIPLES OF MICROBIAL COLONIZATION

The preceding chapters have already described the needs that a microorganism experiences in its quest to survive and proliferate. The establishment of a normal flora on any colonized surface implies that the needs of the microflora can be met at that site. Indeed, as a corollary of this concept, it may be stated that the nutrients available, as well as other physical and chemical characteristics of that environment, determine the types of microorganisms that can inhabit that environment. The unique microenvironments provided by the body's epithelial structures, which are exposed to or communicate with the outside world, provide ecologic niches that are filled by microorganisms. For these microorganisms such tissues represent their natural habitat.

The concept of ecology is a powerful one, and it is an essential first step in understanding numerous infectious disease processes because it helps to provide an understanding of the natural interactions between hosts and microorganisms. To apply ecologic concepts to the microflora of the vagina, it is necessary to identify the environment, its inhabitants, and the interactions that occur between the two. Figure 3-2 outlines the various elements of the vaginal ecosystem.

The end result of the interactions between host and microorganisms is the colonized host. The complexities involved are only partly understood and partly implied. The logical starting place for this discussion is with the host in its germ-free or axenic state. This is the natural state for the fetus, until parturition; an infant becomes colonized within hours after birth. Each germ-free tissue of the newborn presents to microorganisms certain tissue architecture, nutrients, and antimicrobial factors. Each of these host attributes is determined by the genetic heritage of the host and are specifically properties of the host tissue. In addition to the host attributes that are genetically derived, there are homeostatic mechanisms that ensure that the conditions that are attractive to the microorganism will continue throughout the life of the host.

The other half of this ecosystem is the microbial community. It is important to emphasize that most ecosystems, including human tissues colonized by microorganisms, are inhabited by more than one species. The fact that different organisms possess different repertoires of physiologic characteristics allows them to interact with each other on the colonized tissue. The various species that combine to make up the vaginal flora may interact cooperatively through the mechanism of cross-feeding or by segregating on different tissue attachment sites, or they may compete with each other for space and nutrients, thereby balancing the numbers of individual species.

The colonized host consists of the microbial community plus the underlying tissue. However, the interactions of these two components, combined with the influences of factors external to the tissue or external to the host, make an ecosystem that is constantly changing and constantly adjusting to these various forces. Studies with germ-free animals have demonstrated that the colonized tissue as well as the colonized host are altered by the presence of a normal flora. The induction of certain tissue enzymes, the longevity of the host, and the flux of water and solutes across epithelia have been documented as being affected by the presence of the normal flora, although such observations have predominantly involved the gut flora. The way in which the presence of a normal flora affects the vagina has not been documented, although possible effects may involve the rate of

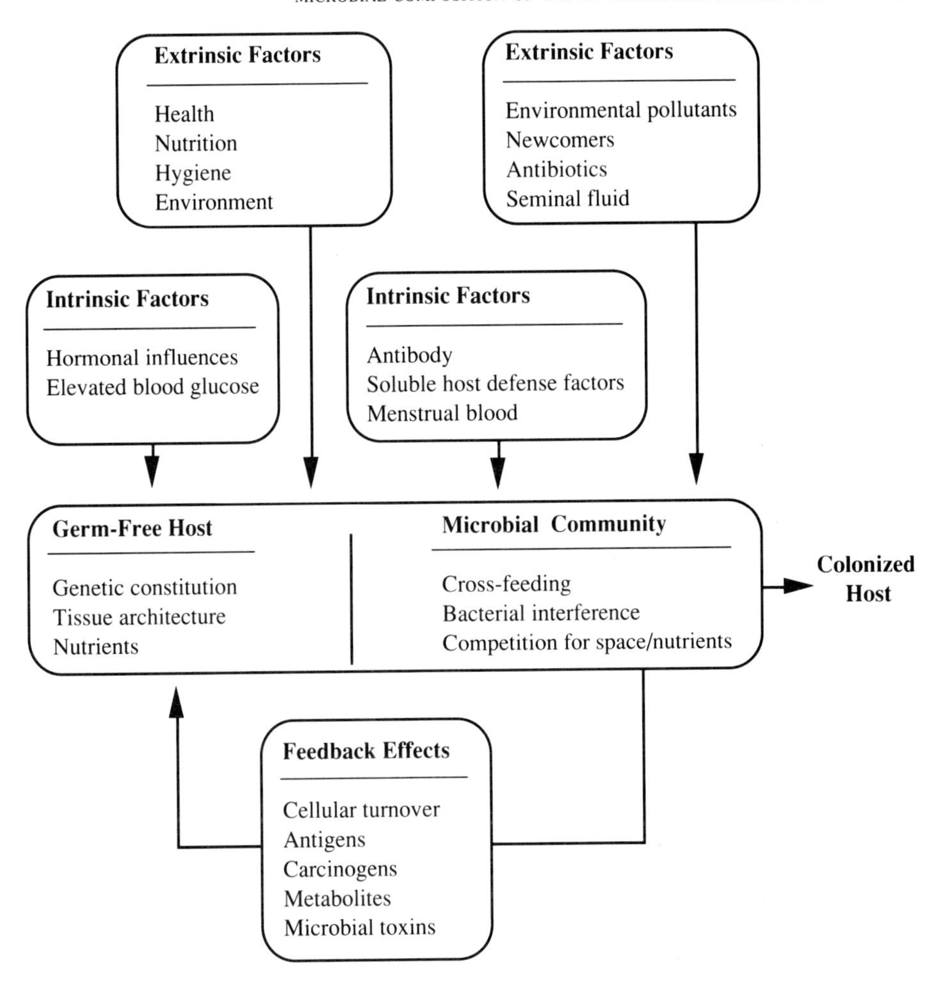

Fig. 3-2. *Composition of the vaginal ecosystem.*

epithelial cell regeneration, recruitment of leukocytes, and presentation of bacterial antigens to local immune systems.

Influences on the microbial flora and on the colonized tissue from within the body are described as intrinsic factors. For example, one of the most profound effects on the epithelium of the vagina comes from the influence of estrogen, which causes the vaginal squamous epithelium to thicken and mature. Animal studies have demonstrated that the tissue changes occurring after estrogen stimulation have a profound effect on the normal flora, causing bacterial numbers to increase more than 10,000-fold. Such effects are primarily manifested in the tissue, and the change in the tissue prompts an alteration of the flora. There are undoubtedly direct intrinsic effects on the flora as well. The production of local antibody may directly influence the composition of the flora, along with other host defense factors such as lysozyme and transferrin in the genital-tract secretions. It has been suggested, but not adequately documented, that additional sugar in the vaginal fluid in women with poorly controlled diabetes mellitus can cause the proliferation of some microbial species.

Extrinsic factors affecting either the microbial inhabitants of the vagina or the vagina itself are also determinants of colonization. The environment in which the host lives and her general state of health and nutrition provide influences that, although unquantifiable, must be recognized as some of the complex factors that specify the nature of the vaginal microenvironment. For example, the fact that cigarette smoke components may be secreted into the cervical mucus emphasizes the possibility that a variety of compounds may be placed in proximity to the vaginal flora despite their origin outside the body. Extrinsic influences that may directly affect the flora could include systemic or topical antibiotics; exogenous microorganisms, which may attempt to proliferate and compete with normal flora organisms; and components of seminal plasma.

This is not an exhaustive summary of the influences that determine the nature of the vaginal flora. However, the list is long enough to make clear the important point that the vaginal flora is sensitive to changes in the tissue it colonizes, as well as sensitive to things that directly influence the growth of microorganisms. Ultimately, these characteristics of this colonized tissue indicate that the vagina and its normal flora represent a dynamic ecosystem. Although the flora is characteristically composed of a limited range of bacterial species in limited numbers, the flora is subject to qualitative and quantitative alterations in response to the influences on this microenvironment.

METHODS OF STUDYING THE GENITAL MICROFLORA

It should be stated at the outset that in a few specific instances the practicing physician may wish to determine whether specific organisms are present in the vaginal flora. However, comprehensive identification of the bacterial species present in the vaginal flora is primarily an academic pursuit and not something that a clinician solely interested in patient care would determine. For the reasons noted above, knowledge of the microbial flora that may be expected to be present in a typical premenopausal woman is important. This information can be used to predict which organisms are likely to be involved in postoperative or postpartum infections and to tailor empirical drug therapy to these organisms. Likewise, the probability of the presence of certain organisms, such as toxic-shock-associated strains of *Staphylococcus aureus* or group B beta-hemolytic streptococcus, is of interest in developing predictions regarding the risks these organisms pose.

Historically, a Gram-stained smear of material obtained from the vaginal pool or from the endocervix has been employed as an indication of the composition of the normal flora. However, given the current state of microbiologic techniques, direct microscopic examination offers little except the ability to detect certain microorganisms that have prominent morphology and are relevant to specific disease processes. When the vaginal flora became a subject of study just before the turn of the century, smears of the vaginal flora were examined microscopically and revealed an abundance of many gram-positive rods. The vaginal lactobacilli were named Döderlein's bacillus, after the pioneering investigator. It became popular to equate a pure flora of Döderlein's bacillus with vaginal salubrity. Culture of asymptomatic women reveals a more complex flora, however, than that suggested by microscopic examination alone.

When investigating the composition of the normal flora, a contemporary investigator will obtain material on a sterile swab from high in the vagina or obtain the sample by rotating a swab in the endocervical canal. Alternatively, the vagina may be lavaged with a small amount of prereduced anaerobic diluent (a salts solution with a reducing agent and an oxidation-reduction indicator), which may be used for quantitation of the flora after appropriate dilution and plating on specific media. Other quantitative methods have included collection of material on a calibrated loop. However, vaginal lavage probably samples a larger area of the vaginal lumen than does the calibrated loop. Hence, it may be argued that lavage samples are more representative.

Regardless of the method of collection, the specimen must be placed in a transport system to protect anaerobic organisms from oxygen and to prevent rapidly growing organisms from proliferating, with the result that isolating minority species becomes difficult or impossible. The microbiologist must receive the specimen and plate it on suitable media as soon as possible. Usually rich media are used for primary isolation to provide for the nutritional needs of the most fastidious of organisms, although selective media may be helpful in enhancing the probability of finding minority species, which are easily obscured. Thus, for complete flora studies, numerous primary culture media may be employed.

The literature is replete with surveys of the vaginal flora. It is clear from these reports that such microbiologic surveys are subject to real as well as artificial differences. Because the vaginal flora is a dynamic entity, the time during the menstrual cycle or the time during a woman's reproductive life when a culture is taken matters with respect to the flora identified. Even if a cohort of women could be found who could be well matched with respect to medical history, demographics, and other criteria, one would still expect to find individual differences in the vaginal flora among these women. Therefore, the purpose of the various available reports on the vaginal flora is to demonstrate the range of organisms present in the lower genital tract and to provide an estimate of the prevalence of specific organisms.

Some differences in the reported prevalence of microorganisms in the female genital tract relate to the adequacy of the techniques used for culture and identification of the organisms. For example, the obligate anaerobic bacteria require special handling, and this fact may account for the vastly different rates of anaerobic isolation reported in the literature. The adequacy of the primary isolation media also determines which organisms can be recovered. Another difference among reports on the composition of the vaginal flora is the frequently changing taxonomic designation of the organisms cultured. Mycoplasmas do not grow on most standard microbiologic media used for general bacteriologic work. Some organisms such as the *Neisseria* species are easily obscured by other organisms and are best recovered from selective media. Thus, the nuances of culture techniques are important to the adequacy of microbial cultivation, although it is often difficult, if not impossible, to ascertain the specific methods of handling cultures in the various reports in the literature. With this caution in mind, the next section will provide a description of the composition of the flora without implying more precision in the data than is warranted.

COMPOSITION OF THE FLORA OF ASYMPTOMATIC WOMEN

The vast majority of research on the normal flora has been directed at describing its composition rather than developing an understanding of why specific organisms colonize this tissue. Consequently, it is possible to describe the normal flora within rather broad limits. An extremely precise definition of the composition of the normal flora is not possible because various investigators have selected their study subjects in different ways and have used different sampling techniques, methods of culture, and methods of identification. The detail with which organisms are identified also varies among studies. The result is that the studies reported in the literature are useful only for gaining a general appreciation of the nature of the flora.

With these limitations in view, it is appropriate to describe three aspects of the vaginal flora: heterogeneity, the prevalence of individual species (qualitative description of the flora), and the number of viable microorganisms present on the vaginal epithelium (quantitative description of the flora).

Heterogeneity of the Vaginal Flora

Historically, much was made of the homogeneity of the vaginal flora, which often is conceived to consist almost solely of lactobacilli. Although the presence of anaerobic bacteria in the human vagina has been known for several decades, the cumbersome nature of anaerobic culture techniques in the past reduced the number of investigations that carefully evaluated both the aerobic and anaerobic microflora. Culture of the microbial flora on enriched medium under aerobic as well as strictly anaerobic conditions provides evidence that the flora in an individual host is composed of numerous species. Investigators using contemporary microbiologic techniques report an average of two to three aerobic species per patient and an average of three to four anaerobic species per patient. In individual subjects, the normal vaginal or endocervical flora may contain 15 bacterial species simultaneously. Although this degree of diversity is greater than the traditional view of the vaginal flora as synonymous with lactobacillus colonization, vaginal flora is certainly less diverse than gut flora, which some investigators estimate may simultaneously contain hundreds of species of microorganisms.

The fact that multiple bacterial species occur in the vaginal and endocervical flora implies that infections resulting from the normal flora will most likely have a polymicrobial etiology. The frequently polymicrobial nature of obstetric and gynecologic infections has been repeatedly demonstrated in such diverse cases as postoperative infections, postpartum endometritis, pelvic inflammatory disease, and a variety of other conditions. This polymicrobial etiology in turn will be shown to have practical significance for therapy.

Qualitative Description of the Normal Flora

A considerable number of specific organisms may occur in the lower genital tract of asymptomatic women, although certain organisms are more likely to be present than others. The vaginal flora commonly contains gram-positive aerobic or facultatively anaerobic rods and cocci and gram-negative facultatively anaerobic rods. The prevalence of species within these groups of organisms is summarized by Fig. 3-3.

Among the gram-positive rods, there is a high prevalence of lactobacilli, a fact known from the earliest observations of the normal flora. These organisms are usually benign and are rarely involved in infections. Indeed, these organisms are often considered to play a protective role in the vagina, preventing colonization or overgrowth of virulent species. The diphtheroids, like the lactobacilli, are generally not speciated and are considered to be benign organisms. They are also commonly isolated members of the normal flora, although they are apparently less prevalent than lactobacilli.

The staphylococci are also commonly isolated from vaginal or cervical cultures of asymptomatic women. The less virulent *Staphylococcus epidermidis* is far more prevalent than *Staphylooccus aureus*, which is significant both as a cause of wound infection and as the etiologic agent of toxic shock syndrome. This does not imply, however, that every *Staphylococcus aureus* isolate is capable of producing toxic shock.

The streptococci are very commonly isolated from vaginal cultures. Some, but not all, species play a significant role in infectious processes. The streptococci are a diverse group of organisms, and commonly the vagina will harbor more than one *Streptococcus* species simultaneously. Figure 3-3 indicates the range of prevalence observed for four groups of streptococci commonly found in vaginal cultures.

Generally, the initial classification of streptococci is made on the basis of the hemolytic reaction of these organisms. The ability of the streptococci to completely lyse sheep red blood cells present in blood agar is referred to as beta-hemolysis. The beta-hemolytic streptococci include some of the most virulent species, although the ability to destroy erythrocytes is not directly responsible for the virulence of these organisms. Two major human pathogens are included among the beta-hemolytic streptococci, and these may be distinguished both on physiologic and serologic grounds. The serologic distinction lies within the C-carbohydrate of the bacterial surface and is relevant not only to the beta-hemolytic streptococci but also to all streptococci. The serogrouping of the streptococci designates most vaginal isolates of beta-hemolytic streptococci as group A (*Streptococcus pyogenes*) or group B (*Streptococcus agalactiae*). The groups are frequently referred to as Lancefield groups in recognition of the pioneering work of Rebecca Lancefield.

Group A streptococci are characterized not only by specific antisera but also by sensitivity to bacitracin, which is used to presumptively identify these organisms. Group A streptococcus, while usually known for its association with sore throat, scarlet fever, and such poststreptococcal sequelae as acute glomerulonephritis and rheumatic fever, can also be present in wound infections and puerperal sepsis. Indeed, in the era before Semmelweis this organism caused the dreaded childbed fever. This organism is an uncommon inhabitant of the genital tract and rarely causes obstetric or gynecologic infections unless transmitted from some exogenous source.

The majority of vaginal isolates of beta-hemolytic streptococci belong to Lancefield's group B. These may be identified by means of either the CAMP test or hydrolysis of hippuric acid. The CAMP test is a procedure in which staphylococci and group B streptococci are grown on the same blood agar plate, resulting in a synergistic hemolysis.

The group B streptococci are primarily of interest as causes of neonatal sepsis, although they may play a role in other infectious processes as well. As indicated by Fig. 3-3, this organism is present in approximately one in six asymptomatic women.

The alpha-hemolytic streptococci produce incomplete hemolysis on blood agar or simply cause a greening around their colonies (which accounts for the name "viridans streptococci") as they alter the hemoglobin of the erythrocytes in the medium. These organisms belong to various species and rarely is there an attempt to speciate them. Most of these organisms are considered to be fairly innocuous, although they may be isolated from the polymicrobial melange found in obstetric and gynecologic infections. The nonhemolytic streptococci also include various species with limited invasive potential. This group of organisms is sometimes designated as gamma-hemolytic, or sometimes gamma streptococci. The latter is to be preferred inasmuch as the term gamma-hemolysis is an oxymoron.

The last group of streptococci of particular interest belongs to Lancefield's group D and contains the enterococci. This group of organisms does not display a uniform hemolytic reaction and may include species that are nonhemolytic. The enterococci are especially hardy organisms and resist the antibacterial effects of bile, elevated temperature, and high salt concentration and are resistant to specific antibiotics to which most other streptococci are susceptible. These organisms are presumptively identified by their growth on agar containing bile salts and by the hydrolysis of esculin. The role of these organisms in infectious processes is controversial. However, because their antibiotic susceptibility patterns may differ from those of the other streptococci, it is important to know whether these organisms must be considered significant when they are present in mixed infections and whether they require specific therapy.

It has only been recognized in recent years that one of the most prevalent of all gram-negative bacilli is *Gardnerella vaginalis*. From the time of its discovery in the 1950s, this organism has been a source of taxonomic confusion. The gram reaction of the organism was uncertain; it has been in the gram-negative genus *Haemophilus* and the gram-positive genus *Corynebacterium*. At the time of its discovery, it was considered to be present as a pathogen in the condition described as "nonspecific vaginitis." The organism typically produces tiny colonies that may be overlooked on some media and may not grow at all on other media. Thus, its presence was not always reported in older studies. Currently, it is recognized that this organism is present in as many as half of asymptomatic women and in an even larger proportion of women with bacterial vaginosis. This organism, while playing a role in bacterial vaginosis, is of low virulence and generally is not associated with infections.

Enteric-type organisms may be present in the vaginal flora. The predominant facultatively anaerobic gram-negative rod is *Escherichia coli*. This organism is part of the normal flora of the bowel as well as the vagina and is the organism most commonly responsible for urinary-tract infection. It participates in postoperative infections and may be involved in intrapartum chorioamnionitis and neonatal sepsis, particularly as a cause of neonatal meningitis. Other gram-negative rods are found less frequently in vaginal cultures from asymptomatic women. These are usually *Enterobacter*, *Klebsiella,* and *Proteus*. Significantly, however, *Pseudomonas* spp. are usually not isolated from the lower genital tract except under specific conditions, such as in immunosuppressed patients, oncology patients, and in those individuals who have received treatment with an antibiotic to which *Pseudomonas* spp. are resistant.

Other groups of microorganisms may be present in the normal flora, including the mycoplasmas. These organisms have no rigid cell wall, but in other respects they are similar to conventional bacteria. *Mycoplasma hominis* and *Ureaplasma urealyticum* may be present in the normal flora but appear to be increasingly prevalent with increasing numbers of sexual partners. Because these organisms require exacting growth conditions and because they cannot be visualized by conventional light microscopic techniques, most general surveys of vaginal flora have excluded them from study. In addition, no clear pathogenic role for these organisms was apparent in the past, which may also have contributed to their neglect. Newer evidence has suggested a role for these organisms in intrapartum infections and possibly in pelvic inflammatory disease.

The presence of a fungal flora should also be noted. *Candida albicans* is the most common inhabitant of the genital flora and is present in approximately 15% of women, although this prevalence increases under some circumstances such as pregnancy. The second most commonly isolated yeast in women is *Candida glabrata*. A very low prevalence of various other yeast species is observed.

The presence of fungus-like bacteria, including members of the genus *Actinomyces*, is occasionally reported in cultures of the female genital tract. These anaerobic microorganisms are often found in association with intrauterine contraceptive devices and raise the question of whether they are part of the normal flora or whether they colonize as the result of the presence of a foreign body. Some information from Scandinavian researchers indicates that actinomycetes may be present far more frequently in the normal vaginal flora than is generally recognized. It is estimated that actinomycetes or related organisms may be present in the cervical flora of 4-25% of intrauterine contraceptive device users.

Although strictly anaerobic bacteria were known to exist in the lower genital tract of asymptomatic women and were also known to be involved in infectious processes, these organisms were cumbersome to isolate. Consequently, the taxonomy of the anaerobes was poorly developed. In the 1960s improvements began in culturing and identifying anaerobic bacteria, which translated into interest in the 1970s in defining the anaerobic flora of the vagina and endocervix. The results of these studies have revealed a great diversity among the anaerobic species that may be found in the female genital tract. The anaerobic flora includes gram-positive rods and cocci and gram-negative rods and cocci.

Table 3-1 provides a summary of the anaerobic microorganisms that have been isolated by investigators who used contemporary methods to evaluate the anaerobic microflora of the lower genital tract. Many of these species are also isolated from polymicrobial infections, but it is difficult, if not impossible, to evaluate the precise significance of each organism to the disease process. The anaerobic gram-positive cocci are usually found in the majority of women. The anaerobic lactobacilli are frequently isolated from asymptomatic subjects as are other apparently innocuous gram-positive nonsporulating rods.

Of major interest to the clinician are members of the genus *Bacteroides* since these organisms appear to be more consistently virulent than many of the other members of the anaerobic normal flora. In particular, *Bacteroides fragilis* possesses a constellation of

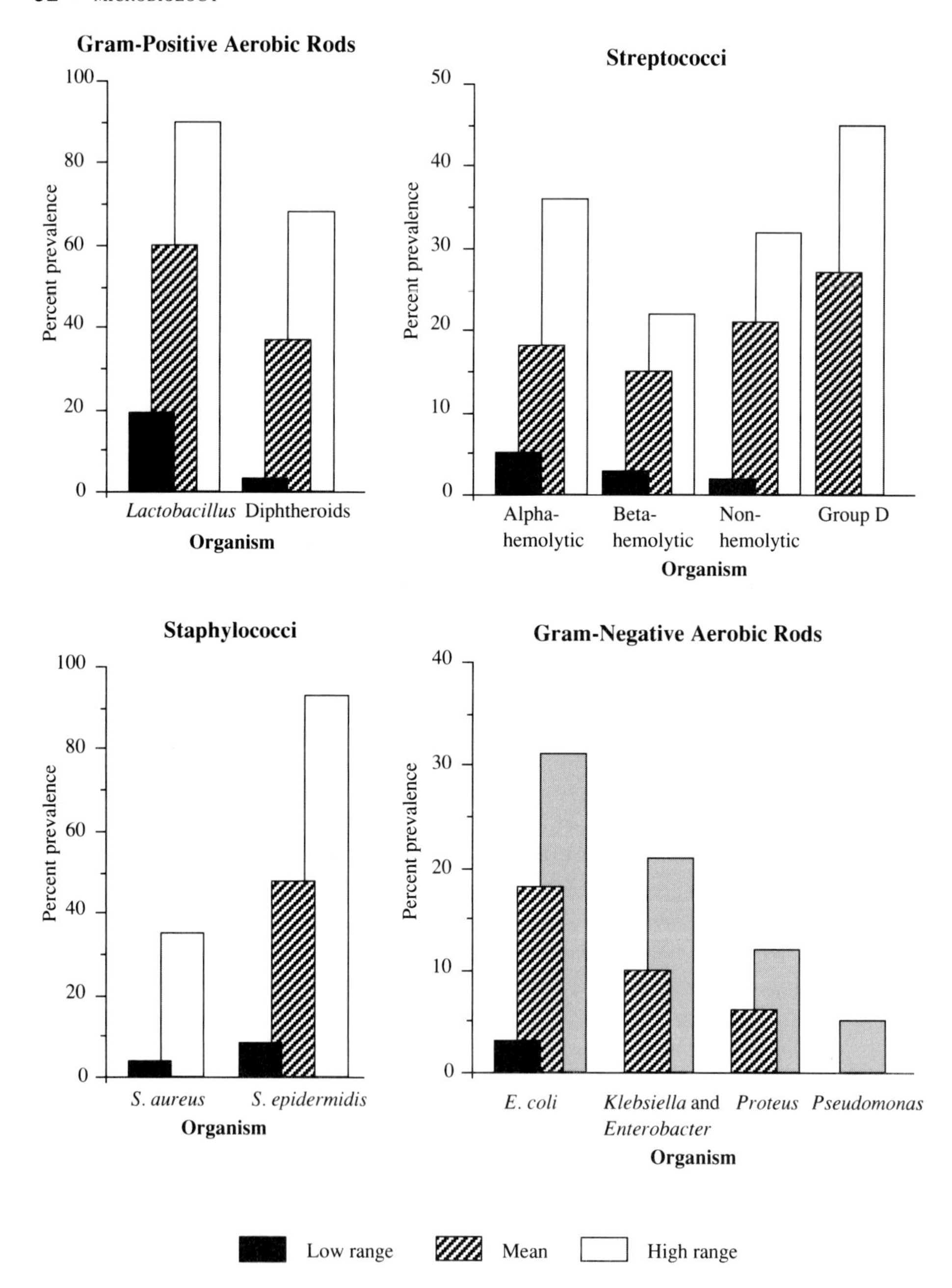

Fig. 3-3. *Prevalence of species within groups of organisms likely to be found in the lower genital tracts of asymptomatic women.*

Table 3-1. *Range of Reported Prevalences for Anaerobic Bacteria Isolated from the Genital Tract of Asymptomatic Women*

Major Characteristics	Genus Name	Range of Prevalence
Gram-positive, nonsporulating rods	*Lactobacillus* sp.	10-52%
	Propionibacterium sp.	0-8%
	Eubacterium sp.	3-31%
	Bifidobacterium sp.	2-10%
Gram-positive, sporulating rods	*Clostridium perfringens*	0-4%
	Clostridium sp.	0-13%
Gram-positive cocci	*Peptococcus* sp.	7-65%
	Peptostreptococcus sp.	15-76%
	Gaffkya sp.	0-2%
Gram-negative rods	*Fusobacterium* sp.	7-28%
	Bacteroides sp.	0-46%
	B. fragilis	0-17%
	B. melaninogenicus	0-33%
	B. bivius	0-34%
	B. disiens	0-16%
Gram-negative cocci	*Veillonella* sp.	9-27%

virulence attributes that include the ability to elaborate exoenzymes including beta-lactamase, proteases, and heparinase; the production of a capsule that is antiphagocytic and promotes abscess formation; and succinic acid as a metabolic by-product. Succinic acid has been implicated in the disruption of the biologic functions of leukocytes. Notably, the prevalence of *Bacteroides fragilis* is probably less than 10%. Although not as virulent, *B. disiens* and *B. bivius* are frequent inhabitants of the female genital tract and are probably more commonly involved in gynecologic infectious processes than *B. fragilis*. Because they produce beta-lactamase, *B. disiens* and *B. bivius* may pose a problem for therapy similar to that associated with *B. fragilis* infections.

Quantitative Descriptions of the Normal Flora

The understanding of the vaginal flora in quantitative terms is less well developed than knowledge of other aspects of the flora. Considerable difficulties are involved in sampling the flora, disaggregating it from sites of tissue attachment, and diluting and plating the samples in such a way that those organisms present in low numbers as well as those present in high numbers will be found. Because of this current state of the microbiologist's art, only a few observations will be made in this section.

Quantitative data are probably important to disease processes and even our perception of the qualitative aspects of the normal flora. If an organism is present in very low concentration in 100% of women, our qualitative culture techniques may miss it in 99% of women. The flora could be far more heterogeneous than we realize. When only qualitative data are reported, organisms present in high concentration and those present in low concentration appear to be of equal importance, a situation that may or may not be true.

Given the interrelationships among virulence of organisms, number of organisms, and host defenses (see chapter 2), it is reasonable to believe that the number of normal flora organisms will have some significance for disease. The size of the bacterial inoculum that occurs during surgical section of the vaginal epithelium depends on the concentration of specific organisms in the flora. Such organisms as *Candida albicans* or *Gardnerella vaginalis*, which may be considered normal flora, may be involved in disease when their numbers are increased.

One of the important observations resulting from quantitative studies of the vaginal flora is that the anaerobic species outnumber the aerobes by a factor of about 10. This preponderance of anaerobic bacteria has also been observed in the gut, but at that site the anaerobic population is about 1,000 times larger than the aerobic population. When the vaginal flora contaminates a surgical site, a significant part of that inoculum will consist of anaerobes. Notably, these organisms tend to grow less rapidly than those organisms that may grow aerobically. The number of anaerobic bacteria present in the vaginal flora may outnumber enteric organisms by a factor of 1,000 or more. Because enteric organisms such as *Escherichia coli* can proliferate rapidly, dramatic alterations in the proportions of various microbial species may occur in the early stages of a mixed infection.

PHYSIOLOGIC INFLUENCES ON THE VAGINAL FLORA

Changes in Flora During Stages of Life

Reference has been made earlier in this chapter to the concept of microecology, which gives rise to the concept that a change in the habitat of the microorganism will be accompanied by a change in the microbial inhabitants of that ecosystem. This concept is well illustrated by the vaginal flora because the tissue changes that occur during the lifetime of women result in documentable alterations in the normal flora.

The vaginal epithelium of female infants is estrogenized as a result of transplacentally acquired estrogen. The tissue contains glycogen and the vaginal pH is acidic even before colonization by acidogenic microorganisms. The vagina of the infant is receptive to colonization by a microflora similar to that of the mother. As the estrogens acquired from the mother disappear from the infant's circulation, the flora tends to be characterized by a lower prevalence of lactobacillus and a higher prevalence of anaerobic species. In post-menopausal women who do not receive estrogen replacement therapy, there is also a tendency toward reduced lactobacillus, diphtheroid, and staphylococcal prevalence compared with that seen in premenopausal women.

At menarche the regular cycles of estrogen secretion and menstruation begin. These in turn influence vaginal colonization. In studies that compared vaginal culture results from the secretory and follicular phases of the menstrual cycle, organisms associated with postoperative sepsis were more prevalent in the first half of the cycle. Posthysterectomy infections occurred in a greater percentage of women who underwent the procedure during the first half of the cycle compared with the second half of the cycle.

Menstruation obviously represents a significant change in the vaginal microenvironment and is attended by documented changes in the flora. A greater diversity of species is observed during menses than after cessation of flow. *Streptococcus faecalis*, staphylococci, and some anaerobic species are more prevalent during menstruation.

Pregnancy also produces physiologic changes that are reflected by changes in the flora. Longitudinal studies throughout pregnancy indicate that a progressive change in the flora occurs, resulting in increasing prevalence of lactobacillus and *Candida* and decreasing prevalence of some anaerobic species.

Other Physiologic Influences on the Flora

Most of the documented influences on vaginal flora have been related to estrogen. It may be anticipated that a variety of other influences may be exerted on the flora by physiologic changes in the host. It is believed that in some women the carbohydrate present in the vaginal pool is increased in cases of poorly controlled diabetes, or that dietary consumption of carbohydrate may appear in the vagina, leading to a greater likelihood of vaginal candidiasis. At present, it is appropriate to note that additional controlled investigation of this problem is needed. Although some physicians may propose dietary restrictions in women with problematic *Candida* infection, such an approach may not lead to uniform success.

Sexual intercourse, particularly accompanied by the deposition of seminal fluid, may also alter the microbial flora of the vagina. Arousal accompanied by increased vaginal lubrication can raise the vaginal pH, and the ejaculum can likewise buffer the pH at a relatively alkaline level. Various components of seminal fluid may affect the flora. Seminal fluid contains high levels of zinc and fructose. It also has been reported that a protein identified as seminalplasmin possesses antibacterial activity. The exact role of this and other seminal fluid components has yet to be established.

NONPHYSIOLOGIC INFLUENCES ON THE VAGINAL FLORA

Certain hygiene habits may have at least theoretic influences on the flora of the external genitalia and possibly the flora of the vagina as well. Many so-called feminine hygiene products are marketed, but little is known about their effect on the genitalia and genital microflora. Chronic use of deodorant sprays, panty shields, and douche products may have ill-defined effects. Studies with antibacterial douches have indicated that vaginal flora is only transiently affected by these products after a single use. Nevertheless, chronic douching may tend to select for organisms that are not susceptible to the active ingredients. Studies of skin flora indicate that occlusive dressings promote the proliferation of bacteria. Especially when used by women with mild urinary incontinence, panty liners may tend to increase the moistness of the vulva, promoting the survival and growth of organisms, particularly those from the perianal area. Despite the lack of concrete information regarding the specific effects of hygiene practices on the vaginal, vulvar, and perianal flora, a physician may occasionally need to discuss these practices with a patient. He or she may suggest discontinuance of the use of products that may theoretically be related to irritation or infection as an empirical contribution to alleviating symptoms.

Because the practice of obstetrics and gynecology involves surgical manipulations of tissues that possess a normal flora, it is appropriate to consider what effects surgical procedures may have on the flora. Most attention in this regard has focused on vaginal and abdominal hysterectomy. The assumption is made that the normal vaginal and cervical flora present at the time of surgery contaminates the operative site, resulting in infection. Before considering the accuracy of this theory about the genesis of postoperative infection, it is appropriate to consider the results of vaginal flora studies that address this issue.

If one cultures the vaginal flora before vaginal or abdominal hysterectomy in a group of women and repeats the culture during the first three to five postoperative days, one discovers that the prevalence of lactobacillus colonization tends to decline significantly. However, the prevalence of aerobic and anaerobic gram-negative rods, which are commonly found in postoperative infections, increases dramatically in the vaginal flora of these women. This tendency persists even when parenteral antimicrobial prophylaxis is employed, although the magnitude of the bacterial proliferation appears to be diminished.

This information modifies the previously mentioned assumption that postoperative infection arises simply as a result of contamination of the operative bed by vaginal flora. More appropriately, the manner in which the flora proliferates after a surgical procedure determines which organisms will contaminate the operative site. As noted in chapter 2, one of the determinants of infection is the number of microorganisms gaining access to the host's tissues. Thus, the presence of organisms that are commonly associated with postoperative infections in the vaginal flora of women undergoing hysterectomy is not necessarily ominous. More important may be the degree to which these species proliferate in the immediate postoperative period.

Longitudinal studies have also indicated that, compared with third-trimester cultures or cultures taken six weeks postpartum, cultures taken on the third postpartum day showed increases in some components of the flora other than the lactobacilli. Striking increases in the prevalence of *E. coli* and *B. fragilis* were observed in cultures taken on the third postpartum day compared with third-trimester cultures.

The reasons why the flora should change during the days immediately after surgery or parturition is unknown. The presence of blood or lochia or the presence of devitalized or traumatized tissues could play a role. Regardless of the specific mechanism involved, it is apparent that these changes probably have implications for the development of infections. As the relationship among the flora changes and the development of infections becomes better understood, the practicing physician may be presented with novel methods for controlling infections.

4

Antibiotics in the Host-Parasite Interaction

This chapter presents basic concepts of antibiotic use that are frequently buried under the daily routine of practical antibiotic use. Bringing to the fore the conceptual framework that underlies antimicrobial use emphasizes the importance of using antibiotics rationally and pragmatically.

Antibiotics are among the most commonly misused drugs, primarily because many of the antibiotics (although certainly not all) possess minimal toxicity to the host, and the physician may take comfort in the idea that even if the drug does no good, it will probably do no harm. Furthermore, because of the daunting number of individual antibiotic products available for parenteral and other use, coupled with the differences in similar products claimed by manufacturers, the clinician may welcome simplification. But oversimplifications must be resisted. The answer lies in familiarity with the principles of antibiotic action and use rather than in overuse or rigid adherence to specific protocols.

The notion that antibiotics are employed to kill bacteria must to be refuted at the outset of this chapter. The use of these important drugs is a bit more subtle. In the previous chapters emphasis has been placed on the role of host-microorganism interactions in the genesis of disease. Host-microorganism interactions must likewise be considered in the treatment of disease.

Disease is an unbalanced interaction between host and microbe. The advantage that the microorganism enjoys over the host results in symptoms. Antibiotics tip the balance in favor of the host either by killing microorganisms or by preventing their unbridled replication, which will permit the host to control the infection by means of its intrinsic defenses. Therefore, in individuals with intact host defenses bacteriostatic antibiotics may provide a therapeutic result comparable to that of a bactericidal antibiotic.

There are three primary requirements of a clinically useful antibiotic:

- It must interfere with some structure or function of the microorganism.
- Toxicity to the microorganism must be achieved without undue toxicity to the host.
- The antibiotic must reach the site of infection in sufficient concentration to cause an antimicrobial effect.

INHIBITION OF MICROORGANISMS

Antimicrobial drugs are able to interfere with specific bacterial processes or structures or the biosynthesis of critical structures. In the following paragraphs the antimicrobial mechanisms of action of specific classes of drugs will be summarized. However, this is not an exhaustive listing of all antimicrobial agents available for clinical use. Rather. it is limited to those compounds most commonly used by the obstetrician-gynecologist. The antibacterial spectrum of activity and mechanisms of resistance are briefly summarized as well.

Inhibition of Cellular Metabolism

FOLATE BIOSYNTHESIS

Among the earliest available antimicrobial agents were the sulfonamides, which take advantage of the fact that prokaryotic organisms synthesize folate from a *p*-aminobenzoic acid (PABA) precursor. Folic acid is a cofactor involved in single carbon transfers and among its critical functions is thymidine synthesis. The sulfonamides are structural analogs of PABA and become incorporated into an inactive folate analog.

The terminal stage in the folate pathway is shared by prokaryotes and eukaryotes. Dihydrofolate is obtained in bacteria by the condensation of PABA with other precursors as described above, whereas the human host obtains dihydrofolate from dietary sources. In both cases dihydrofolate is reduced to its active form, tetrahydrofolate. The drug trimethoprim blocks the activity of the reductase enzyme and is useful as a therapeutic agent because the bacterial dihydrofolate reductase enzyme is five orders of magnitude more susceptible to the effect of trimethoprim than is the human form of the enzyme.

The form of these drugs used in current therapy is a combination of trimethoprim and sulfamethoxazole. The theoretical reason for the combination is to achieve antimicrobial synergy by blocking two different sites in the same biochemical pathway. However, there is some dispute as to whether sulfamethoxazole plays a significant role in vivo in this combined dosage form.

The inhibitors of the folate pathway are intrinsically active against a variety of bacteria including *Neisseria*, *Haemophilus*, *Staphylococcus*, *Streptococcus*, and most *Enterobacteriaceae*. The sulfonamides are not active against anaerobic bacteria, *Streptococcus faecalis*, or *Pseudomonas* sp. Trimethoprim is active against *Chlamydia* and some parasites including *Pneumocystis carinii*.

Resistance to the sulfonamides is usually plasmid mediated, and simultaneous resistance to trimethoprim may sometimes be present on the same plasmid. Plasmid-mediated resistance to trimethoprim results from the production of an alternative dihydrofolate reductase that is less sensitive to the inhibitory action of trimethoprim. The resistance of *Pseudomonas* to trimethoprim, however, results from the impermeability of this organism to the drug.

ELECTRON TRANSPORT

Metronidazole depends on its interaction with ferredoxins of the anaerobic electron transport chain. Metronidazole and other nitroimidazoles are taken up by susceptible bacterial cells. In the intracellular environment the nitro group undergoes a series of reductions that presumably involve the addition of single electrons. Oxidation-reduction reactions involving single electrons can generate highly reactive radicals. It is proposed that a nitroso free radical or superoxide generated by the reoxidation of a reduced nitro group is the principal toxic product of this reaction. These reactive intermediates then react with and cause damage to the bacterial chromosome and other susceptible cellular targets. The intracellular reaction of metronidazole results in the scission of the molecule into two inactive end products, acetamide and 2-hydroxyethyl oxamic acid.

Because oxygen can accept electrons from the ferrichromes, it may compete with the action of metronidazole. Thus, metronidazole is active against anaerobic microorganisms and also requires anaerobiosis for its full antimicrobial effect.

Metronidazole has a spectrum of activity that is particularly useful for gynecologic practice. This drug is effective against the anaerobic protozoa including *Entamoeba*, *Giardia*, and *Trichomonas*. It also has broad-spectrum activity against anaerobic bacterial species, although minimum inhibitory concentration (MIC) values tend to be lower for gram-negative anaerobes than for gram-positive nonsporulating anaerobes. Anaerobic and microaerophilic streptococci are resistant. *Gardnerella* and *Campylobacter* species, which are microaerophilic, are susceptible. Interestingly, some antibacterial activity against facultative organisms such as *Escherichia coli*, *Proteus*, and *Klebsiella* has been noted when tested under anaerobic conditions.

Mechanisms of drug resistance are not well understood for metronidazole. Most available evidence points to an alteration in the ability of the resistant organism to reduce the drug intracellularly. This may result from a decreased ability of the oxidoreductases to use metronidazole as an electron acceptor.

STEROID BIOSYNTHESIS

The triazole drugs—including miconazole, ketoconazole, fluconazole, butoconazole, and itraconazole—are active against fungal pathogens due to their interference with the synthesis of steroids, which are not present in bacteria but are structural entities in the membranes of yeast and filamentous fungi. The site of action of these compounds appears to involve the activity of cytochrome P-450 in the fungal microsomes. The precursor of ergosterol, lanosterol, has a methyl group at the 14 position on the sterol nucleus that is oxidized by the microsomal cytochrome. The 14α-methylated sterols disturb the integrity of the cellular membrane and possibly membrane-associated enzyme systems such as chitin synthetase. Cells that have been exposed to triazole drugs show a variety of morphologic aberrations.

Inhibition of Nucleic Acid Replication

As noted previously, the geometric rearrangement of the entwined coils of DNA is essential to bacterial replication. This function is perturbed by the quinolone drug nalidixic acid, and since its introduction numerous other fluorinated congeners have been developed. Notably, the enzyme affected by the quinolones is a topoisomerase enzyme, DNA gyrase, named for its ability to insert negative supercoils into bacterial DNA. The effect of the quinolones is selective for the bacterial enzyme; mammalian cells are resistant to quinolone action.

The mechanism of quinolone action involves more than inhibition of DNA replication. A variety of cellular processes also involve the selective transcription of specific segments of the DNA as part of the normal process of gene regulation and expression. As a consequence, the quinolones have a broad range of activities with respect to cellular function, which may include suppressing the expression of virulence factors. These drugs display a broad spectrum of antimicrobial activity against various bacterial species because the same DNA gyrase is essential to virtually all types of pathogenic bacteria.

Resistance to the action of quinolone antimicrobial agents is the result of spontaneous mutation of the genes coding for the gyrase enzyme. It is believed that two simultaneous mutations are required to render the gyrase enzyme insusceptible to the action of a given quinolone drug, which results in an exceedingly low rate of spontaneously appearing resistance. In most instances, resistance to one quinolone drug does not result in cross-resistance to other congeners.

Resistance caused by plasmids does not seem to occur, which may be due in part to the ability of the drug to inhibit the replication of plasmid DNA as it does chromosomal DNA. Indeed, it is believed that the quinolones may cure bacterial populations of plasmids; the quinolones are known to reduce the expression of plasmid-specified beta-lactamase.

Inhibition of Protein Synthesis

The cellular process of making proteins as specified by the genetic code is a complex phenomenon involving structures (ribosomal components) as well as various enzymes and cofactors. Translation of the RNA-encoded information by the transport of individual amino acids carried by transfer RNA provides numerous opportunities for disruption of the process of producing functional proteins. Therefore, it is not surprising that a large number of available antibiotics are inhibitors of some aspect of protein synthesis.

AMINOGLYCOSIDES

This class of compounds is structurally related to streptomycin, the first available drug in this class and the progenitor of numerous compounds that are or have been clinically useful, including kanamycin, gentamicin, neomycin, tobramycin, and amikacin.

Although in theory the cessation of protein synthesis should not be lethal for bacteria, the aminoglycosides interfere with protein synthesis and are bactericidal. In contrast, other protein synthesis inhibitors are bacteriostatic. Until recently, the bactericidal action of the aminoglycosides was enigmatic. The action of the aminoglycosides resides primarily in interaction with the small ribosomal subunit, where the drugs cause an unfaithful reading of the RNA messenger resulting in the production of defective proteins. It is now believed that these proteins are dysfunctional and upon incorporation into the membrane disrupt its function. The resulting irreversible damage to the membrane is considered to be involved in the lethal effect of the aminoglycosides.

The most significant therapeutic contribution of the aminoglycosides is in controlling infections caused by gram-negative enteric pathogens. Of particular note is the inclusion of *Pseudomonas aeruginosa* within the aminoglycoside spectrum of activity. However, the patterns of susceptibility for individual organisms may vary. In practice, the aminoglycosides may be used in combination with beta-lactam drugs, which may produce an additive or synergistic effect. It must also be noted that anaerobic organisms are intrinsically resistant to aminoglycosides because transport into the cell involves an oxygen-dependent respiration-linked process.

During the history of aminoglycoside use, there have been occasional outbreaks of resistant *E. coli*, *Proteus*, *Pseudomonas*, and *Klebsiella*. There have also been examples of progressively increasing prevalence of aminoglycoside resistance among isolates of gram-negative organisms in hospitals where one of the aminoglycosides has been used extensively.

The mechanisms of resistance include blocked transport of the drug into the cell, a ribosomal alteration resulting in decreased affinity for aminoglycoside, and plasmid-mediated drug-altering enzymes. There are at least 11 phosphorylating or acetylating enzymes that inactivate aminoglycosides. Gentamicin and tobramycin are substrates for eight of the inactivating enzymes, whereas amikacin is a substrate for only two, which may account for the fact that organisms resistant to gentamicin, tobramycin, and netilmicin are susceptible to amikacin.

TETRACYCLINES

Since their introduction into clinical practice in the late 1940s, the tetracyclines have become some of the most heavily prescribed antibiotics. Currently, most uses by obstetrician- gynecologists involve two semisynthetic derivatives, doxycycline and minocycline. This antibiotic class is characterized by a bacteriostatic mechanism of action resulting from the drug's interaction with the 30S ribosomal subunit or perhaps to some degree with the 50S subunit or the complete ribosome. The precise site of action remains obscure. However, the concept of faulty aminoacyl RNA binding to the 30S subunit, with the resulting failure in initiation complex formation, is favored by many as an explanation.

The tetracyclines are considered to be broad-spectrum antimicrobial agents and are intrinsically active against many gram-positive and gram-negative aerobic and anaerobic species, although specific resistance patterns may be variable. *Proteus*, *Pseudomonas*, and *Serratia* are intrinsically resistant. In addition, some other classes of organisms are inhibited, including chlamydia, rickettsia, mycoplasma, and spirochetes.

The greater activity, improved pharmacologic properties, and decreased bacterial resistance of the "second-generation" semisynthetic tetracyclines (doxycycline and minocycline) have resulted in a wider potential use for these compounds.

Resistance to tetracyclines may occur by a plasmid-mediated mechanism that does not serve to alter the drug. The resistance genes in gram-negative bacteria are frequently associated with transposon 10, which may also be associated with other R factors. The resistant cells appear to respond to the presence of tetracycline by the production of at least three proteins for which a specific function is unknown. Although susceptible bacteria apparently take up tetracycline by an energy-dependent transmembrane transport, resistant cells appear to exclude the drug and may even actively pump it out of the cytoplasm.

MACROLIDES

Erythromycin is the most widely used of the macrocyclic antibiotics and exerts a bacteriostatic action on 70S ribosomes by binding to the large subunit. When peptidyl transferase activity is selectively extracted from ribosomes, their susceptibility to erythromycin is removed. The erythromycin-binding site involves interaction with at least two ribosomal proteins, both of which are involved in the peptidyl transfer reaction. A practical aspect of this mechanism of action is that erythromycin is antagonized by other antibiotics such as chloramphenicol that affect the peptidyl transfer reaction. The exact mechanism of erythromycin activity may involve the blocking attachment of peptidyl tRNA to the ribosome, because ribosomes that are in the process of peptide chain elongation are refractory to erythromycin.

The antimicrobial spectrum of erythromycin includes primarily gram-positive facultative organisms and a variety of other organisms, including *Neisseria gonorrhoeae*, *Chlamydia trachomatis*, *Haemophilus ducreyi*, *Treponema pallidum*, and *Mycoplasma* sp. Some of the obligately anaerobic species are also within the spectrum of erythromycin activity, although *Bacteroides fragilis* strains are generally insensitive.

A plasmid-borne resistance mechanism that is induced by the presence of erythromycin has been identified in *Staphylococcus aureus*. The product of this plasmid is a methylase that methylates a structural RNA present in the 50S ribosome, rendering it insensitive to the action of erythromycin, as well as of lincomycin. A second mechanism of resistance involves the production of altered forms of two structural proteins present in the 50S subunit.

CHLORAMPHENICOL

This compound is a simple, low-molecular-weight antimicrobial agent that is uncommonly used because of its reputation for causing idiosyncratic aplastic anemia. Despite this, some specific indications exist, although these are generally unrelated to the obstetric and gynecologic specialty. However, there is renewed interest in the drug as an alternative for anaerobic infections involving *Bacteroides*. The drug binds to and inhibits protein synthesis on 70S bacterial and mammalian mitochondrial ribosomes. The binding is reversible, involves sites on the 50S subunit, and is antagonized by erythromycin or lincomycin. Not all details of the inhibitory process are understood, but it does seem clear that some aspect of the peptidyl transferase reaction is affected by chloramphenicol.

Two mechanisms of resistance to chloramphenicol are known. Among aerobic organisms, resistance is due to a plasmid-specified acetyl transferase that causes mono- or diacetylation of the drug, resulting in inactivation of the compound. It is believed that the acetylation occurs at position 1 of the molecule. A nonenzymatic isomerization takes place, moving the acetyl group to position 3, which leaves position 1 available to again be acetylated, resulting in 1,3-diacetoxychloramphenicol. The mechanism among anaerobic bacteria involves the reductive degradation of the paranitro group of this compound with consequent loss of antimicrobial activity.

CLINDAMYCIN

This low-molecular-weight drug is the chlorinated derivative of lincomycin. This inhibitor of protein synthesis, which is structurally different from erythromycin and chloramphenicol, has a mechanism of action similar to these drugs. It binds to the 50S subunit of 70S ribosomes and interferes with the peptidyl transferase reaction.

The spectrum of activity of this drug is interesting in that gram-positive organisms are inhibited, whereas gram-negative, enteric-type organisms are not. This difference is caused not by variations in cellular permeability but rather by an intrinsic difference in the ribosomes. Clindamycin is able to bind to gram-positive but not gram-negative ribosomes. However, the drug is active against both gram-positive and nonenteric gram-negative anaerobic organisms, including *Bacteroides fragilis*.

In some cases resistance to clindamycin involves a phenomenon described as dissociated resistance, in which erythromycin-resistant strains of *Staphylococcus aureus* remained sensitive to lincosamides until exposure to erythromycin and then became cross-resistant to both drugs. This appears to be related to the previously described phenomenon of ribosomal RNA methylation and has occurred in streptococci and pneumococci. Some methicillin-resistant staphylococci are cross-resistant to clindamycin. Upon repeated and prolonged exposure to the drug, staphylococci may develop gradual stepwise resistance to clindamycin.

Inhibition of Cell Wall Biosynthesis

The cell wall is a unique structure in the bacterial cell. Theoretically, agents that damage that structure should have little toxicity for the mammalian host. The bacterial cell wall is exceedingly complex. It requires a complicated biosynthetic system, which manufactures a portion of the new cell wall material within the bacterial cell and then transports it outside the cell where the building blocks are inserted into the growing cell wall macrostructure.

A substantial number of antimicrobial agents have been discovered that interfere with various aspects of cell wall biosynthesis. However, many of these do not have recognized application in the obstetric and gynecologic specialty and will not be discussed with respect to their specific modes of action.

BETA-LACTAM ANTIBIOTICS

These compounds, which include the penicillins and cephalosporins, are among the most useful and most widely prescribed antimicrobial agents not only in obstetrics and gynecology, but in almost all specialties. The mode of antibacterial action relates to the terminal stages of cell wall synthesis, in which the precursors of cell wall peptidoglycan become incorporated into the growing wall structure and cross-linking of the linear arrays of peptidoglycan material add strength to the structure. The cross-bridging of the cell wall material occurs with the aid of at least two enzymatic reactions. One of these is a transpeptidase reaction that does not require ATP for bond formation and is responsible for linking the pentaglycine to an adjacent acceptor molecule. The second reaction involves a DD carboxypeptidase that removes the terminal D alanine from a site adjacent to the cross-link. Beta-lactam antibiotics may be structurally analogous to the D-alanyl-D-alanine complex that is so vitally involved at the cross-linking site, and they appear to interfere with both of the reactions in gram-positive cells.

The ultimate destruction of bacteria exposed to beta-lactam antibiotics appears to involve more than just the weakening of the cell wall due to the absence of cross-linking. An endogenous bacterial enzyme described as autolysin is responsible for opening spaces in the existing peptidoglycan to allow insertion of new material and to allow the separation of daughter cells. When penicillin affects bacterial cells, the endogenous autolysin may act in a more unrestrained manner, further weakening cell wall structure. Because cell growth is not inhibited by beta-lactam antibiotics, cells may continue to increase in size. Bizarre morphologic shapes such as surface blebs or filamentous forms may occur, and ultimately physical limitations result in lysis of the affected cells.

Gram-negative cells interact with beta-lactam antibiotics in a more complicated manner than is seen with gram-positive cells. In *Escherichia coli*, six penicillin-binding proteins have been identified that have transpeptidase or carboxypeptidase functions that may be concerned with lateral wall or transverse septum formation, control of cell shape, and attachment of lipoproteins to the peptidoglycan. The various beta-lactam antibiotics display differences in binding to these penicillin-binding proteins and as a consequence produce different ultrastructural lesions on the gram-negative cells.

Because the beta-lactam antibiotics have been so useful, many derivatives have been made available for clinical use. Because each beta-lactam antibiotic has its own particular spectrum of activity, it is not appropriate to attempt to describe a spectrum of activity that applies to the entire class of antimicrobial agents, and no attempt will be made here to summarize the antibacterial spectrum associated with specific beta-lactam agents. Penicillin has a spectrum limited to gram-positive organisms, although it also is effective against *Neisseria gonorrhoeae* and *Treponema pallidum*. Penicillins such as ampicillin and cephalosporins such as cephalothin that have gram-negative coverage were developed, but they lacked activity against *Pseudomonas*. Penicillins with specific chemical modifications, including carbenicillin, ticarcillin, and the acylureido penicillins such as azlocillin and mezlocillin, are now available, and they provide antipseudomonal activity. Others have been developed that are resistant to degradation by beta-lactamase enzymes, including the penicillins methicillin and oxacillin and the second- and third-generation cephalosporins. Anaerobic coverage has also been sought. Although many anaerobic species are susceptible to penicillin, some newer cephalosporins and penicillins have been introduced specifically for coverage of anaerobic species, including *Bacteroides fragilis*.

Because of the considerable number of penicillins and cephalosporins that have been approved for therapeutic use, there is frequently confusion about terms used in regard to these compounds. The need to deal with the vast amount of information about these drugs has given birth to a classification scheme in which the cephalosporins are described as belonging to the first, second, or third generation. Unfortunately, there is some ambiguity inherent in these categories because spectrum of activity, chemistry, and date of development have all played a part in classification. Despite the fact that concrete rules of taxonomy do not exist for the cephalosporins, the usefulness of this imperfect scheme lies in the ability to anticipate similar clinical utility of drugs belonging to the same generation. The reader is cautioned, however, that individual therapeutic agents possess unique attributes such as protein binding, toxicity, pharmacodynamics, dosage schedule, specific spectrum of activity, and cost, all of which affect a decision for clinical use in a particular patient.

The first generation of cephalosporins is exemplified by cephalothin. First-generation agents are mainly active against cocci, with the exception of the enterococci, and are also active against some gram-negative rods. They are resistant to the extracellular beta-lactamases produced by staphylococci, but because of the heterogeneity of beta-lactamases, bacterial genera such as *Bacteroides*, *Enterobacter*, *Serratia*, and *Pseudomonas* remain outside their spectrum of activity. The second generation of cephalosporins includes compounds that are cephamycins rather than true cephalosporins. The cephamycins have a methoxy group substituted at position 7 of the 7-amino cephalosporinic acid nucleus. As a result, this group of drugs, which includes cefoxitin and cefotetan, has the ability to encompass *Bacteroides* in their spectrum of activity, although this is at the expense of some loss of efficacy against the gram-positive cocci. The third-generation cephalosporins have enhanced the gram-negative spectrum of activity, which may include *Pseudomonas*, *Proteus*, and *Serratia*.

The primary mechanism of resistance to the action of beta-lactam antibiotics is the production of penicillinase or cephalosporinase, which have the ability to hydrolyze the beta-lactam ring. There is substantial diversity of action among beta-lactamases, which may be plasmid mediated or chromosomal, inducible or constitutive, secreted as an exoenzyme or retained in the periplasmic space. As mentioned above, a class of penicillins (eg, methicillin, oxacillin) has been developed that resist the action of penicillinases, but because they still lack certain pharmacologic features and have a limited spectrum of activity, they do not represent the ultimate all-purpose antibiotic. Another approach to the problem of beta-lactamases is the discovery of beta-lactamase inhibitors such as clavulanic acid and sulbactam. When these are combined with a beta-lactam antibiotic, the antibiotic molecules are protected from enzymatic degradation and the apparent spectrum of activity is expanded.

In addition to beta-lactamase-mediated resistance, methicillin resistance in staphylococci has been recognized as an emerging problem. This resistance is apparently not due to chemical modification of the antibiotic by the organism, but rather to some change in the cell surface structure that makes it impermeable to the drug.

VANCOMYCIN

Current clinical use of this drug is relatively limited. It is active only against gram-positive organisms. Its mechanism of action is derived from its ability to bind to membrane-bound D-alanyl-D-alanine; that is, while the peptidoglycan precursors are in the process of transit from the intracellular site of synthesis to the extracellular site of incorporation into the growing cell wall. Because the spectrum of activity includes *Clostridium difficile*, *Staphylococcus aureus*, and methicillin-resistant staphylococci, vancomycin is primarily used to treat conditions involving these organisms. Resistance appears to be chromosomally associated and appears slowly in a stepwise fashion.

Inhibition of Cell Membrane Integrity

Because the cell membranes of humans and bacteria have many similarities, many antimicrobial agents that affect membrane integrity demonstrate toxicity that limits their clinical utility. One class of membrane-altering drugs has relevance for the obstetrician-gynecologist and will be briefly discussed.

The drugs nystatin (employed topically) and amphotericin B (used parenterally) have the ability to form complexes with sterols that are found in the membranes of yeast and fungi. Because bacteria do not have sterols in their membranes, they are refractory to these drugs. Because mammalian membranes contain sterols, they are potential targets of the polyene drugs. However, amphotericin B and nystatin have a selective affinity for ergosterol and a lesser affinity for cholesterol. Because the former is present in yeast and fungi, they are most affected by these drugs. Nevertheless, the risk of toxicity still exists, especially with amphotericin B.

SELECTIVE TOXICITY

The concept of selective toxicity was conceived by Ehrlich even before antibiotics were a reality. Selective toxicity refers to the fact that antibiotics are relatively more toxic to microorganisms than to the host. As suggested by the preceding discussion, the specific sites of antimicrobial action commonly involve some structure or function that is either unique to the microorganism or sufficiently different in microorganisms compared with humans to spare the patient untoward effects. Despite the wide margin that exists between the therapeutic and toxic doses of antimicrobial drugs, complete freedom from adverse effects of these agents has never been achieved. The following discussion summarizes some of the most important drug toxicities observed among the antimicrobial agents typically used in the obstetric and gynecologic specialty.

Concerns of the Obstetrician-Gynecologist

The obstetrician-gynecologist is faced with several concerns about drug toxicity. These are concerns that apply to nonantibiotic as well as antibiotic drugs. First, one must consider the untoward effects in the uncompromised adult patient. Second, those patients with impaired renal or liver function may be subject to toxicities related to an inability of the body to eliminate the drug. Third, if the patient is pregnant, one must be aware of the intrauterine life with its particular susceptibility to teratogenic influences, especially during the earliest stages of development. This raises issues of both drug toxicity and transplacental transfer. Finally, the secretion of drugs in the milk and colostrum of nursing mothers may pose a risk of toxicity for the infant. The abbreviated summary that follows is an overview and should not be used in lieu of complete familiarity with published reports of toxicity and dosage standards.

Toxicity of Antimicrobial Agents

TRIMETHOPRIM AND SULFAS

These drugs are associated with undesired effects that occur in a small percentage of treated individuals. An estimated 3% of treated individuals experience allergic reactions to sulfonamide drugs. A slightly smaller percentage of individuals experience eosinophilia, and very rarely a hemolytic crisis occurs in association with glucose-6-phosphatase deficiency, a situation that may likewise involve nitrofurantoin or nalidixic acid. In breast-fed newborns a theoretical risk exists of jaundice due to the displacement of bilirubin from serum-binding sites by sulfonamide drugs, although the amount of sulfonamide in the milk is probably sufficiently low to make this complication unlikely. In contrast, trimethoprim is associated with fewer adverse effects than sulfonamides. In individuals who are at the borderline of folate deficiency, a situation that may apply to some pregnant women, trimethoprim may antagonize folate uptake. This problem can be obviated by treating the patient with folates, which does not interfere with the action of trimethoprim because bacteria cannot store the folate.

METRONIDAZOLE

Because of its action on DNA, therapy with metronidazole is avoided during pregnancy (particularly the first trimester) unless its use is urgently required. This dictum is not based on concrete evidence of teratogenicity or mutagenicity, but rather represents caution in the absence of data. In adult patients, very high doses of the drug given intravenously are associated with neurologic sequelae. Metronidazole may produce a disulfiram-like reaction as well.

TRIZOLES

Trizole antifungal drugs are usually administered to the gynecologic patient in topical form, and absorption is limited. The use of these agents during the first trimester of pregnancy is avoided because there is some absorption of the drug from the vagina. Local irritation is probably the most common problem. Orally administered ketoconazole, used for systemic fungal infections, bears a risk of hepatoxicity.

QUINOLONES

Quinolones, especially the newer agents, are a class of antimicrobial drugs that will probably enjoy widespread use. However, at present, limited use has produced limited knowledge of adverse effects in the clinical setting. A very good safety record has been obtained with the quinolones, although animal studies have indicated that because of injury to joints in juvenile animals, the drugs should be avoided during pregnancy and lactation and likewise should not be given to pediatric patients.

AMINOGLYCOSIDES

Aminoglycosides can cause ototoxicity due to the effects on the eighth cranial nerve, as well as renal toxicity. Tobramycin and amikacin appear to be less toxic to the kidneys than gentamicin. Adjusted dosages are required for patients with renal compromise, and serum concentrations are monitored to maintain drug concentrations below toxic levels. All aminoglycosides cross the placenta; streptomycin has been incriminated as causing adverse effects on the fetus, but there are reports of gentamicin use during pregnancy without fetal damage. However, these reports are not synonymous with proof of safety. Generally, therapy with these drugs is avoided during pregnancy. A final, albeit rare, complication of aminoglycoside use is neuromuscular blockade, which can occur in patients with myasthenia gravis or hypocalcemia and may be fatal. Other neuromuscular-blocking agents may act in synergy with aminoglycosides when used concomitantly.

TETRACYCLINES

Tetracyclines are well known for their affinity for the calcium-containing tissues, including teeth and bones. Oxytetracycline causes the least staining of dentition. Tetracyclines should not be used in pregnancy or in children under the age of 8 years. This drug has a reputation for predisposing patients to oral or vaginal yeast infections. Hepatotoxicity, including liver failure in pregnant women, has occurred with intravenous administration of tetracycline, although this route of administration is rarely necessary.

MACROLIDES

Macrolides, as exemplified by erythromycin, are beset with the tendency to produce gastrointestinal symptoms, including epigastric pain, nausea, vomiting, and diarrhea when oral preparations are used. These symptoms are generally self-limiting and respond to discontinuance of therapy with the drug. Perhaps one of the most significant secondary problems is the failure of patients to complete a course of therapy as a result of gastrointestinal symptoms. Erythromycin use during pregnancy has a very good safety record, although nausea leading to patient noncompliance may be particularly worrisome during pregnancy. A serious but rare complication of erythromycin use is cholestatic hepatitis, which appears to be related to a hypersensitivity reaction. The disease is reversible if therapy with the drug is discontinued early.

CHLORAMPHENICOL

Chloramphenicol toxicity is surrounded by significant controversy. The idiosyncratic aplastic anemia that occurs in approximately 1 of 25,000-40,000 patients treated is both serious and rare, making risk-benefit assessment difficult. Most recorded cases of stem cell aplasia were associated with oral chloramphenicol. Therefore, if chloramphenicol use is limited to parenteral forms, the risk of aplastic anemia is anticipated to be small. Because it is impossible to predict which patients will succumb to this complication, the drug is reserved for a limited number of indications. In addition to aplastic anemia, chloramphenicol is associated with a reversible form of bone marrow depression. The "gray baby syndrome" may occur in infants who receive large doses of the drug. This complication may result from the inability of the immature liver to conjugate and dispose of the drug, resulting in high concentrations that cause inhibition of mitochondrial electron transport. A constellation of symptoms are associated with this complication, but ultimately cyanosis, shock, and death may occur within 24-48 hours of the onset of symptoms.

CLINDAMYCIN

Clindamycin is noted for causing pseudomembranous enterocolitis, which may occur during antibiotic therapy. This diarrheal disease is accompanied by cramps, fever, and dehydration. The condition may resolve spontaneously or progress in some individuals to toxic megacolon. The condition is related to the suppression of the normal bowel flora, with the subsequent outgrowth of *Clostridium difficile*, which produces two toxins. Toxin A causes a secretory diarrhea and induces a hemorrhagic inflammatory reaction in experimental animals. Toxin B is a cytopathic toxin for tissue culture cells. Discontinuation of therapy with the offending drug and treatment with oral vancomycin are effective in these cases. Finally, it should be emphasized that many oral or parenteral antibiotics have the ability to cause antibiotic-associated enterocolitis.

BETA-LACTAMS

Hypersensitivity reactions are the most notable untoward effect of beta-lactam antibiotics. Penicillin allergy is widespread within the population and may be manifested in various forms, including serious immediate reactions such as anaphylaxis, due mainly to IgE associated with mast cells and basophilic cells. The reaction may occur within minutes and can lead to death if proper intervention is not undertaken. Hypersensitivity due to circulating immune complexes and complement may be responsible for urticarial rashes, which arise during the first few days after treatment and represent the most common form of allergic reaction to penicillin. Delayed reactions include a condition resembling serum sickness. A hemolytic anemia may arise as a result of penicillin combining with erythrocytes to form an antigen complex that can be attacked and lysed by antibody and complement.

Penicillin-allergic patients given cephalosporins are subject to hypersensitivity reactions. The cross-reactivity between penicillins and cephalosporins is not complete. However, a history of penicillin allergy should elicit caution regarding any intended cephalosporin use. Sensitivity to cephalosporins also can occur in the absence of penicillin allergy.

Various other adverse reactions have been recorded for the numerous cephalosporin antibiotics. Of note is the occurrence of bleeding problems associated with cephalosporins that are substituted with the *N*-methyl thiotetrazole (NMTT). NMTT may stimulate liver microsomal enzymes to activate clotting factors II, VII, IX, and X, causing hypoprothrombinemia. The carboxylation of glutamic acid residues on the clotting-factor precursors provides sites for calcium binding and is a vitamin K-dependent process. As a consequence, either vitamin K depletion or carboxylase enzyme inhibition may interfere with the essential posttranslational modification of the clotting factors. NMTT may inhibit the carboxylase, but only after cleavage from the antibiotic molecule. When NMTT dimerizes via a disulfide bridge, it becomes a more potent inhibitor of the carboxylase. A further consequence of NMTT appears to directly involve the availability of vitamin K cofactors needed for the carboxylation. Vitamin KH2 is oxidized to the 2-3 epoxide during the decarboxylation step and must be regenerated by an epoxide reductase. NMTT inhibits the regeneration of the reduced vitamin, depriving the system of this cofactor.

VANCOMYCIN

Vancomycin is a relatively safe antibiotic. When it was first available, contaminants that copurified with the active compound were responsible for most of the adverse effects. The most serious adverse effect of this drug is eighth-cranial-nerve toxicity leading to tinnitus, high-frequency hearing loss, and deafness, although the effect is dose related. If serum concentrations remain below 80-100 μg/ml, ototoxicity is unlikely. Allergic reactions may occur but in most cases may be obviated by slow infusion.

POLYENES

Polyene antibiotics, exemplified by amphotericin B, may show a very small margin between therapeutic and toxic doses. Indeed, a majority of patients will not tolerate the full intravenous dose without such undesired effects as fever, chills, anorexia, headache, nausea and vomiting, and generalized pain involving the muscles and joints. Renal toxicity is also common with this compound. Topical or nonabsorbed oral nystatin does not display the toxicity that systemic amphotericin B elicits.

EFFECTIVE CONCENTRATION AT INFECTED SITE

The third requirement for successful antimicrobial therapy involves various host and microbial factors. The pharmacokinetic and pharmacodynamic properties of the antimicrobial agents determine how much drug reaches the infected site and how long it will persist. The toxicity discussed above emphasizes the existence of upper limits on drug dosages. The intrinsic activity of the antimicrobial agent toward the offending microorganism determines the amount of drug that must reach the infected site to produce the desired result. In the following discussion, the laboratory measurements that address these issues will be presented.

Laboratory Tests and Chemotherapeutic Choices

One practical aspect of the diversity of microorganisms is differences in susceptibility to antimicrobial drugs. Differences among species is to be expected, but various strains within a species can also differ in their susceptibility to a particular antibiotic. This fact necessitates the occasional use of laboratory tests to determine an organism's susceptibility to antimicrobial agents.

Because laboratory tests cost both time and money, it is appropriate to use them when necessary and dispense with them when they are not needed. Fortunately, the susceptibility of certain microbial species is predictable, as is the case with *Treponema pallidum*, which is exquisitely sensitive to penicillin and has not demonstrated any propensity to develop resistance. The clinical solution for organisms of this type is to simply identify the microorganism and treat the infection according to a known pattern of susceptibility. In contrast, the majority of gram-negative rods of enteric origin, which figure significantly in surgical infections, are not so predictable in terms of antimicrobial susceptibility and require laboratory testing. Many clinical laboratories routinely include susceptibility testing along with the identification scheme for those organisms, which typically have an unpredictable susceptibility pattern.

Although the term "susceptibility" is used frequently in clinical practice, it is appropriate for a chapter emphasizing basic principles of chemotherapy to discuss the concept of antimicrobial susceptibility. An organism is susceptible to an antibiotic if the antibiotic inhibits the organism at concentrations that may reasonably be achieved in the host. Thus, susceptibility involves considerations of toxicity as well as the attainable concentration of the drug within the host. Susceptibility is therefore defined in terms of practicality.

Laboratory measurements can provide information on the susceptibility of organisms isolated from clinical infections. However, it must be remembered that an infection caused by an organism that is reported as susceptible to a particular antibiotic will not necessarily respond to that antibiotic; the specific laboratory tests do not predict what will happen in the patient. In the laboratory, dilutions of the antibiotic are added to cultures of the bacterium of interest. After a suitable period of incubation, the presence of visible growth is noted in the cultures. The smallest concentration of antibiotic that prevents visible growth is taken to be the MIC. In general, if drug concentrations can be achieved that exceed the MIC by a factor of two to four, the organism is considered to be susceptible to the antibiotic.

In some cases the minimal bactericidal concentration (MBC) is measured. This is the smallest amount of drug required for complete killing of the test organism. This measurement is primarily useful in identifying a condition of antibiotic tolerance in which the MBC is significantly greater than the MIC. This may occur in the case of beta-lactam drugs that lack the autolysins (see "Beta-Lactams" in this chapter) that are responsible for the final lysis of penicillin-treated cells. The practical implication of this phenomenon is that a bactericidal antibiotic like penicillin may behave as a bacteriostatic drug, and the offending organism may recommence growth when therapy is stopped.

For the sake of completeness, the concept of minimal antimicrobial concentration will be mentioned. It is known that various antibiotics may impair the normal function of microorganisms at concentrations below the MIC. These impairments may include retarding some metabolic activities; producing aberrant, albeit viable, morphologic forms; or the suppression of the synthesis of virulence factors. In the host with intact defenses, these suboptimal concentrations of the drug may be of benefit and may explain why, on occasion, a patient improves while receiving an antibiotic regimen that the laboratory report indicates is marginal. The maxim that one must treat the patient and not the culture report is derived from this phenomenon.

The logical question of why susceptibility tests are used at all, given the facts that they will not guarantee the success of a particular antibiotic in the patient and that therapy usually must begin before susceptibility tests are completed, deserves some comment. In the case of the patient receiving therapy, if the response to the antibiotic is deemed inadequate, the susceptibility data may aid in selecting alternative therapy. In addition to guiding the therapy of an individual patient, the aggregation of hospitalwide or nationwide susceptibility data can help in selecting empirical therapy in a rational manner.

The use of aggregated susceptibility data invokes the term "break point." Break points 50, 90, and 95 refer to the MICs required to inhibit 50%, 90%, and 95% of a substantial number of clinical isolates of a certain species of microorganism. Break point data can be used to predict whether a drug given according to recommended dosages has a high or low degree of probability of inhibiting a certain species of microorganism.

In some situations, it is not sufficient to know that a microorganism is susceptible to an antibiotic by in vitro testing. As stated, susceptibility refers to inhibition by concentrations of drug that can be achieved in the serum. But in some cases the serum concentration may be less than expected. It is not always possible to predict when serum concentrations are too low, and, as will be discussed later, inadequate clinical response is not always the result of inadequate antibiotic concentrations. Likewise, problems of toxicity may arise when the antibiotic reaches concentrations that are too great. The ability to measure serum concentrations in these situations becomes important. Serum concentration rarely needs to be measured in ordinary clinical practice except with aminoglycoside therapy. A variety of assay methods are available to determine the concentration of antibiotic in serum or plasma. These may be chemical, immunochemical, or bioassay methods. The availability of such tests varies among clinical laboratories. Such data are essential in establishing the pharmacologic characteristics of new drugs and consequently are usually obtained in research settings. However, in routine hospital laboratories the availability of these tests may be limited.

Pharmacologic Considerations

The pharmacodynamic and pharmacokinetic properties of antimicrobial drugs have a significant impact on the ability to obtain an adequate concentration of drug at the site of infection. The concept of susceptibility has been linked to serum concentrations; however, in most cases of antibiotic use in obstetric and gynecologic practice, the serum is not the site where the antibiotic will act. A substantial portion of the drug given to the patient is unavailable at any time after administration. A portion is associated with serum-binding sites, and part of the free drug is distributed to the infected tissue. Part of the administered dose is also removed by the usual routes of elimination, either renal or hepatic or both. As a practical consequence, serum concentrations of renally excreted antibiotics will typically be poor predictors of efficacy in urinary-tract infection because very high urinary levels may be obtained. Furthermore, the serum concentration and consequently tissue concentrations will be higher immediately after the drug is given and lower before the next dose is administered.

These pharmacologic considerations may have minimal practical importance for drugs that achieve serum and tissue concentrations well above the MIC. However, for those drugs that have attainable concentrations in the host near the MIC, pharmacokinetic and pharmacodynamic behavior exerts a substantial influence. Generally, serum concentrations are reported as the highest concentrations attained. In theory, the fluctuation of antibiotic concentration in vivo should have the most profound consequences when bacteriostatic drugs are used because by definition organisms may recommence growth when the drug concentration declines sufficiently.

During gestation and early in the puerperium, the physiologic accommodations to pregnancy may affect the pharmacokinetics of antimicrobial agents. The expansion in plasma volume and the increased glomerular filtration rate may result in lower antibiotic concentrations. In women given beta-lactam antibiotics in the immediate postpartum period, the drugs may only achieve half their normal serum concentrations. This may necessitate more frequent dosing of certain antibiotics if adequate therapeutic response is not achieved.

Therapeutic Failures

Despite the seemingly best choice of antimicrobial agents in clinical infections, the apparent ineffectiveness of the drug as manifest by a patient's worsening clinical picture is always a possibility. It is essential to ask whether such an occurrence is the result of the use of the wrong drug or if other factors are responsible for the unacceptable clinical outcome. Below is a summary of some the factors that should be considered in attempting to understand therapeutic failures.

- *Treatment begun too late:* In some clinical situations the pathology associated with the infection is not reversed simply by the destruction of the microorganisms, and damage to the host may be the result of toxic products of the microorganism. Microbial replication may have proceeded too far to be arrested by antimicrobial therapy. Some antibiotics are less effective when the bacterial inoculum is large, a situation that may exist in the case of extensive bacterial growth.
- *Insufficient concentration at site of infection:* Various reasons for insufficient drug concentrations are possible. In addition to the problem of the pregnant patient mentioned above, the dosage form, distribution into specific tissues, and vascularity of tissue may be incriminated.
- *Additional procedures required:* Surgical intervention may be required for adequate resolution of an infectious process. Necrotic tissue or accumulations of pus or serous drainage represent sites where bacterial proliferation is promoted, and

because of the lack of vascular supply, antibiotic delivery to such sites may be relatively poor. Removal of retained products of conception, drainage of abscesses, and debridement of necrotic tissues exemplify the types of surgical procedures that may be required as adjuncts to medical therapy.

- *Bacteria in a dormant state:* As described earlier, bacteria typically grow and reproduce at the fastest rate permissible under existing environmental conditions. In some clinical situations, such as in a collection of pus, bacteria may be present in high concentration in a milieu that contains large amounts of metabolites that inhibit growth. Antibiotics that act on bacteria during growth or metabolism will display reduced effectiveness against organisms that have ceased to grow.
- *Superinfection, overgrowth, infection elsewhere:* Even if an antibiotic is effective against the offending organism, a second unrelated infection may be preexistent or arise during the course of therapy. Typically, this infection will not be susceptible to the current therapeutic drug. Common examples include antibiotic-associated colitis or yeast overgrowth in patients receiving tetracyclines.
- *Erroneous diagnosis:* If symptoms of a viral illness are attributed to a bacterium, antibiotic therapy will obviously not be efficacious. Likewise, attributing symptoms of an antibiotic-resistant bacterium to one that is normally susceptible will not benefit the patient.
- *Impaired host defense:* A variety of causes of impaired host defense are known, but most current attention is focused on the acquired immune deficiency syndrome (AIDS). In patients with impaired defenses, bacteriostatic antibiotics may be less effective than bactericidal drugs because the former rely on the host defense mechanisms to contribute to the effectiveness of the drug.

In summary, the principles that guide antimicrobial therapy involve all of the complexities of interactions among the factors of host physiology and defense, microbial metabolism and biosynthesis, and the pharmacologic features of the drug. No formula exists that guarantees the success of a particular antibiotic in a given clinical situation. Thus, rational use of the available drug formulations is the best means of dealing with the infectious complications encountered in obstetrics and gynecology.

Part II
Endogenous Infections

5

Endogenous Gynecologic Infections

VAGINAL CANDIDIASIS

Overview

A variety of species of *Candida* may be found in the vaginal flora, but the predominant species is *Candida albicans*. Likewise, this organism is responsible for the majority of yeast vaginitis cases. About 1 patient in 20 may have yeast vaginitis that involves *Candida glabrata* (known previously as *Torulopsis glabrata*) and occasionally other species, including *Candida tropicalis*, *Candida pseudotropicalis*, or *Candida stellatoidea*. The most virulent as well as most prevalent of these yeasts is *C. albicans*. Most of the following discussion will apply to this organism unless otherwise stated.

Source

Most occurrences of symptomatic yeast vaginitis are presumed to arise from a reaction to organisms indigenous to the vagina. However, other sites of colonization exist, including the mouth and intestines. In a very high percentage of women with vulvovaginal candidiasis, intestinal carriage has been documented. These sites may be more significant for persistence and recurrence of infection than as the source of primary infection. Male colonization leading to sexually transmitted yeast vaginitis is a debatable issue, but it may be involved in some apparent treatment failures.

Microorganism

Yeast cells are larger than bacteria and ovoid in shape. When Gram stained, the cells appear to be gram positive. In contrast to bacteria, yeast cells divide by budding, and in clinical specimens pseudohyphae characterized by elongated cells attached end to end may be noted.

Several specific characteristics of this organism are known to be associated with virulence, but the way in which they actually enhance invasiveness is not clear. Attachment to the vaginal epithelium is an important characteristic and appears to have some relationship to the ability to form germ tubes and pseudohyphae. Superficial invasion of epithelial surfaces is a common feature of mucocutaneous disease. Factors associated with invasiveness include the ability of the organism to exist in a filamentous form rather than in the yeast form. Exoenzymes including phospholipase and proteases have also been named as contributing to the pathogenicity of this organism. Because *Candida* sp. may be found in the normal flora without causing symptoms, future work in the field will undoubtedly attempt to delineate environmental factors that elicit the production of known or suspected virulence attributes in these organisms. Some yeast strains also produce factors that may suppress phagocytic or immune function and may inhibit bacterial growth.

Clinical Features

Mucocutaneous candidiasis involving the vaginal and possibly the vulvar epithelium accounts for most of the *Candida* infections seen by the obstetrician-gynecologist. The typical curdy white discharge accompanied by pruritus and itching is well known, although the character of the discharge may vary and therefore should not be considered a sine qua non of candidiasis. Frequently the vaginal symptoms are accompanied by vulvar erythema and swelling of the labia minora. The exanthem may also extend to the perianal area and in some cases to the inner aspect of the thighs.

Mucocutaneous lesions may also occur in the mouth and throat, and babies can acquire the organism from the birth canal of women who carry the organism. The prevalence of neonatal thrush is far below the prevalence of maternal candidiasis during pregnancy.

Intestinal overgrowth and disseminated candidiasis usually are associated with underlying immunosuppression or endocrine disorder; cutaneous candidiasis is typically associated with macerated and chronically wet skin. These conditions are mentioned for completeness and rarely involve the obstetrician-gynecologist.

Diagnostic Considerations

Because the symptoms of vaginal candidiasis are fairly typical and in recurrent episodes the patient is good at self-diagnosis, the physician frequently succumbs to the temptation not to examine the patient. If the patient is examined, an opportunity is provided to ensure that the diagnosis is correct, to rule out other vaginal conditions, and in the case of recurrent infection to investigate the possibility of other underlying predisposing diseases.

Each patient deserves a physical examination and a microscopic examination of the discharge prepared both as a wet mount and as KOH preparation. The KOH preparation is made by adding a drop of 10% KOH to a sample of the discharge on a slide and briefly warming the slide over a flame. This process digests epithelial cells, accentuating the fungal elements. No staining is needed, and observation with the high-dry objective is sufficient, although reducing the light on the microscope stage by adjusting the condenser lens often provides more contrast between the fungal structures and the background. The microscopically demonstrated presence of budding yeast and pseudohyphae coupled with patient symptoms are usually adequate to establish the diagnosis in an office setting.

If there is a desire to culture the organism, commercially available Biggy agar (Nickerson medium) is easy to use and produces typical brown colonies in 1-3 days. The germ tube test is the laboratory method usually employed to establish that the organism is *C. albicans*. This organism germinates, producing microscopically visualized tubes that extend from the yeast cells when grown in the presence of serum for 3-5 hours at 37°C.

Predisposing Factors

Pregnancy is well known for its association with vaginal candidiasis. The rate of prevalence of positive vaginal cultures among pregnant women is about 30%, compared with about 15% among nonpregnant women. The predilection for candidal symptoms in pregnancy probably is related to the same influences that increase colonization.

Diabetes mellitus that is poorly controlled is also a predisposing factor for vaginal candidiasis. Notably, the presence of yeast infection also diminishes the probability of bringing the patient's blood glucose under control. Other endocrinopathies may predispose patients to candidiasis.

As noted above, cutaneous candidiasis is associated with moist skin. Therefore, it is not surprising that tight-fitting clothing has been blamed for perineal and vulvar symptoms. Although some dispute the scientific validity of this claim, it bears consideration as part of the overall management of the patient with a vaginal infection.

In the past oral contraceptive products were thought to predispose patients to yeast vaginitis. Some observations indicated that estrogenic compounds affect the growth of yeast, and recently estrogen receptors have been identified in yeast. These receptors may have accounted for the growth stimulation. An alternative hypothesis was that estrogen indirectly stimulated yeast growth by enhancing the deposition of glycogen in the vaginal epithelium. Regardless of the mechanism, the currently prescribed oral contraceptive products appear less likely than earlier products to predispose patients to yeast vaginitis, possibly due to lower estrogenic content.

Complications

Serious complications related to vaginal yeast colonization or vaginitis are extremely rare. Nevertheless, intraamniotic infection, chorioamnionitis, and funisitis do occur. According to some who have been diligent in identifying cases, these complications occur more often than they are recognized.

The complication of oral thrush has been mentioned and is surprisingly rare given the high prevalence of the organism in the vaginal flora of pregnant women.

Penile colonization and balanitis may result from sexual intercourse with an individual who has a vaginal infection, although transient penile colonization is more common than infection. Male colonization may be more important in causing persistence of the disease in women than in causing primary infections in men and probably requires consideration mainly in those cases in which relapse of vaginitis occurs after therapy.

Disseminated candidiasis that is associated with immunosuppressed states probably does not arise as frequently from vaginal colonization as it does from gut colonization. The majority of women with vaginal candidiasis also have the organism in gastrointestinal flora. The organism is ubiquitous, and nonhuman locations, including pet animals, are also potential sources.

BACTERIAL VAGINOSIS

Overview

The current understanding of this condition is that it is abnormal colonization involving members of the host's own flora. The vaginal symptoms were once described as nonspecific vaginitis, but the work of Gardner in the 1950s began to suggest that it was a monoetiologic infection caused by *Haemophilus vaginalis*. This organism has now been renamed *Gardnerella vaginalis*, and the perception of its role in the vagina has been drastically altered. The organism is considered to be part of the normal flora in a significant percentage of women, and the condition once called nonspecific vaginitis is called bacterial vaginosis to indicate that the process is not invasive and is not properly considered to be an infection in the classical sense. *Gardnerella vaginalis* increases in number along with obligately anaerobic species and is accompanied by vaginal discharge, altered pH, complaints of malodor, and decreased numbers of lactobacilli.

Source

Current understanding of bacterial vaginosis indicates that both *Gardnerella vaginalis* and the anaerobic species, including *Bacteroides* species and *Peptostreptococcus* species, that are involved in this condition are normal inhabitants of the lower genital tract. Recently, obligately anaerobic motile, curved rods that have been placed into the genus *Mobiluncus* have also been found in some cases of bacterial vaginosis. The dogma concerning the role of this organism is still being developed, but it appears that the function of this organism in bacterial vaginosis may be the same as that of other anaerobic bacteria. It is not yet clear whether there is an exogenous source for *Mobiluncus* species that are present in this condition. Perhaps, like the other predominant organisms, it is simply a resident species that is present in low concentration until vaginosis supervenes.

Microorganisms

A description of the specific virulence attributes of the organisms found in bacterial vaginosis and how they contribute to the clinical symptoms is probably moot. Rather, it is appropriate to consider what ecologic interactions occur among species and how these result in the numerical dominance of the anaerobic bacteria and *Gardnerella* organisms. The microenvironment of the vagina in bacterial vaginosis cases is usually characterized by a pH greater than 5 and a low oxidation reduction potential. There are numerically fewer lactobacilli present and a variety of amines, including trimethylamine, histamine, putrescine, and cadaverine. A theory of how these conditions are produced by the bacteria involved has been set forth and includes a role for both the *Gardnerella* organisms and the anaerobic bacteria.

The work done at Seattle by Chen and his colleagues suggests that elevated pH is a significant factor in bacterial vaginosis. The change in pH may be the result or the cause of this condition. In either case, the elevated pH appears to favor the growth of *Gardnerella* and its production of acetate and pyruvate. The generation of extracellular proteases that release amino acids from tissue sources may also be the result of increased bacterial growth. The anaerobic bacteria may then be stimulated by the availability of such substrates as pyruvate and acetate. In the course of anaerobic growth, the amino acids may be deaminated. These amines may cause tissue reaction, odor, and maintenance of alkalinization of the vaginal milieu. A further result of the growth of such anaerobes as *Bacteroides* is the production of succinic acid, and an increase in the ratio of succinic acid to lactic acid in the vaginal pool has been considered indicative of bacterial vaginosis. Moreover, succinic acid has been shown to inhibit some aspects of phagocytosis, as mentioned in chapter 2, and may further contribute to the colonization pattern seen in bacterial vaginosis.

The above-proposed mechanism of dominance of *Gardnerella* and anaerobic bacteria requires further investigation to establish whether the details of this condition conform to the current theory. In addition, the factors that trigger the change from the normal pattern of colonization to the abnormal pattern of colonization that characterizes bacterial vaginosis need to be elucidated.

Clinical Features

The symptoms associated with bacterial vaginosis may include the presence of increased discharge. This discharge is characterized by a homogenous grayish to milky appearance that does not seem purulent, nor does it have the greenish or yellowish appearance often seen in cases of trichomoniasis. The patient may notice a fishy odor that is exacerbated after intercourse, when the vaginal secretions have been alkalinized by seminal fluid. Likewise, when the physician makes a KOH preparation, the addition of the KOH liberates the fishy odor. As noted above, the pH of the vaginal discharge is usually above 4.5, which may be detected with pH indicator paper. Culture for diagnosis of bacterial vaginosis is usually only applicable to research situations and may require quantitation of the flora, as well as species identification. Gram stain that shows more than 30 gram-negative rods per high-power field and fewer than 5 gram-positive rods that have morphology consistent with lactobacillus per high-power field is indicative of bacterial vaginosis. Wet-mount preparation reveals clue cells, which are vaginal squames covered with bacteria to such a density that the cellular margins are obscured. Immunofluorescent techniques have recently been used to demonstrate that large numbers of *Gardnerella vaginalis* are among the adherent bacteria on clue cells. Despite the distinguishing features of this condition, it should be noted that not all patients are aware of symptoms.

Complications

The fact that the vaginal flora alterations characteristic of bacterial vaginosis may occur without symptoms noted by the patient may raise the question of how significant this condition is. However, there is a more sinister aspect to bacterial vaginosis that relates to this condition's association with more serious problems. Among the concerns for the obstetrician is the statistical link made between bacterial vaginosis and prematurity. Much remains to be done to identify specific mechanisms that may be involved in the precipitation of premature labor by specific bacteria. When such work is done, it may show that the altered flora of bacterial vaginosis is not the primary cause. Nevertheless, the statistical association appears well founded.

The relative increase in gram-negative anaerobic rods may increase the potential inoculum of virulent organisms in other obstetric conditions, including chorioamnionitis, intraamniotic bacterial infection, and postpartum endometritis. Likewise, the complications of upper-tract invasion by gram-negative anaerobic bacteria as part of the process of pelvic inflammatory disease may be enhanced by the increased prevalence of these organisms in patients with bacterial vaginosis. As will be emphasized frequently, the composition of the genital flora determines the organisms that infect contiguous structures, a fact that underscores the potential significance of the flora changes that accompany bacterial vaginosis.

TOXIC SHOCK SYNDROME

Overview

In 1978, Todd described a staphylococcus-related illness in children as toxic shock syndrome (TSS). This severe febrile disease was characterized by hypotensive shock, a sunburn-like erythroderma, and multiple organ involvement. In 1980 a new dimension to this illness was added when TSS in menstruating women was recognized and fatalities were recorded. A propensity for the disease to recur and its association with vaginal tampon use created vast public attention to this illness.

The Centers for Disease Control in collaboration with physicians and epidemiologists who had been at the leading edge of finding new cases developed a stringent set of case criteria to aid in the investigation of the disease and began collecting reports of definite and probable cases. It became apparent that nonmenstrual cases also occurred in men and women, although in the early history of the condition the majority of cases were associated with menstruation.

Currently, tampon users know more about the syndrome, which has probably served to reduce the prevalence of menstrually associated cases. A lower case-fatality rate than that observed in the early years of the epidemic may be the result of physicians' experience with the disease. Much is known about the microorganisms involved in this syndrome and the toxic products that relate to the illness.

Microorganism

Staphylococci possess a large constellation of virulence factors that contribute to the classical staphylococcal infections. However, in the study of strains associated with TSS, it was found that these organisms appear to display a particular set of biologic properties. Although *Staphylococcus aureus* was named for its golden pigment, many clinical isolates are nonpigmented. On primary isolation, however, TSS strains tend to be pigmented. Phage typing places most of the strains in group I. The strains are associated with types 52 and 29, and a significant percentage are nontypical. A high degree of penicillin resistance has been noted, but multiple antibiotic resistance is not typical. Resistance to cadmium and arsenic is common.

The ability of TSS strains to produce certain toxins may depend on the growth conditions used in individual experiments, and this fact must be remembered in evaluating our current knowledge. TSS strains may produce less alpha toxin than other strains; there may also be a shift in production favoring delta toxin over alpha toxin in TSS organisms. Alpha toxin is potent and has numerous biologic effects that will not be detailed here. It is possible that the result of diminished alpha toxin production is that the TSS strains may cause less severe reactions when they infect the host tissues. In addition, the hemolytic reaction on primary isolation media may be diminished among TSS strains compared with non-TSS strains. TSS strains also produce less lipase and nuclease but more casein and hemoglobin proteases. Although these various factors are associated with TSS strains, one should not assume that they necessarily have relevance for the illness.

Of all the toxins, it appears that the one most closely and probably directly related to TSS is now called TSST-1 (toxic shock syndrome toxin 1). This transposon-specified protein toxin has been called by other names in the past, and the differences in nomenclature, methods of purification, and biologic actions have spawned controversy. With the nomenclature settled, it is now possible to state more definitively the actions of this toxin. TSST-1 is pyrogenic and may possess the ability to cross the blood-brain barrier, resulting in a direct effect on the hypothalamus and indirectly causing fever by the release of interleukin-1. Other lymphokines may also be released by the action of TSST-1. Some studies have demonstrated animal lethality and the potentiation of endotoxin effects in rabbits. Cellular systems may be affected by the toxic effects on neutrophil function, reticuloendothelial blockade, and immunosuppression. As noted already, these biologic activities may account for the clinical observations in patients with TSS, but proof of this hypothesis will require further study.

Source

Staphylococcus aureus is not limited only to human hosts. Survival outside the body and colonization of the families of the index case patients with identical strains of staphylococcus is possible. Nevertheless, most cases of TSS have been attributed to endogenous sources. The organism is normally present in the vagina, on the labia or other skin, or on the anterior nares of some asymptomatic individuals, and as a consequence these areas may be the source of the infection. In the menstruating woman who acquires TSS, the usual source is believed to be the vaginal flora. The vaginal flora may also be the source in surgical or postpartum cases of TSS. It should not be assumed, however, that the staphylococci that colonize these sites are necessarily TSS-associated strains. Male colonization of skin, urethra, and nares also occurs, and a portion of the currently noted cases of TSS are associated with surgical procedures in men.

Clinical Features

The majority of TSS cases have occurred among young and previously healthy individuals. The clinical picture begins with rapid onset of fever, nausea, vomiting, and

Toxic shock syndrome case definition

Fever (temperature of >38.9°C)

Rash (diffuse macular erythroderma)

Desquamation, 1-2 weeks after onset of illness, particularly of palms and soles

Hypotension (systolic blood pressure of <90 mm Hg for adults or < 5th percentile by age for children <16 years or orthostatic syncope)

Involvement of three or more of the following organ systems:

- Gastrointestinal (vomiting or diarrhea at onset of illness)
- Muscular (severe myalgia or creatine phosphokinase level that is more than twice the normal limit)
- Mucous membrane (vaginal, oropharyngeal, or conjunctival hyperemia)
- Renal (blood urea nitrogen level or creatinine level that is ≥ twice the normal limit or ≥ five white blood cells per high-power field—in the absence of a urinary tract infection)
- Hepatic (total bilirubin, serum glutamic oxaloacetic transaminase level or serum glutamic pyruvic transaminase level that is more than twice the normal limit)
- Hematologic (≤ 100,000 platelets/μL)
- Central nervous system (disorientation or alterations in consciousness without focal neurologic signs when fever and hypotension are absent)

Negative results on the following tests, if obtained:

- Blood, throat, or cerebrospinal fluid cultures
- Serologic tests for Rocky Mountain spotted fever, leptospirosis, or measles

diarrhea, which may be accompanied by abdominal cramps. Myalgias and generalized weakness occur early in the illness. Within the first 2 days after onset, hypotensive changes leading to shock may develop. On the first through fourth days the typical macular exanthem may develop on the trunk, moving to the neck and extremities; simultaneously, mucous membranes and conjunctiva become inflamed. One to two weeks after onset, desquamation of the palms and soles occurs.

The acute illness is likewise characterized by multisystem involvement that may include, in addition to the above-mentioned myalgias and gastrointestinal symptoms, renal, hepatic, or central nervous system effects. The accompanying box summarizes the features that characterize a definite case of TSS according to criteria established by the Centers for Disease Control. It should be emphasized that not all patients fit this case description with exactitude, and the clinician should not neglect considering TSS in the differential diagnosis and should look for a potential site of infection.

Predisposing Factors

The prevalence of vaginal colonization by staphylococci and certain other organisms is increased both during menses and in the first postpartum week. Studies have also indicated that a shift in the flora also occurs after hysterectomy, which may also enhance the colonization by staphylococci.

Because of the magnitude of attention focused on the association between TSS and menstruation and TSS and tampon use, it is easy to believe that tampons are causally related to TSS that occurs during menstruation. The exact relationship and mechanism remain matters that can be debated on both biologic and epidemiologic grounds. Certainly, an epidemiologic correlation has been made between tampon use and TSS. However, many aspects of this problem have been addressed without a final consensus on the mechanism of this association. Some suggest that tampons act as a sink for growth and toxin production by staphylococci. The role of toxin production and control of mineral availability by the tampon material has been considered, and the relationship between tampon composition and bacterial growth and toxin-producing capability has been studied and debated. Another consideration involves the way in which the tampons are used. If tampons are left in place for a prolonged time, the accumulation of toxin may be presumed. However, tampons were used in vast numbers before the TSS outbreak of 1980, a fact that argues against a simple and direct causal relationship between tampon use and TSS. Nevertheless, in cases of menstrually associated toxic shock, discontinuing use of tampons should be a part of the management.

In contrast to the situation in the early 1980s, TSS is currently not predominantly associated with menstruation, although menstrually associated cases are still being reported. A significant proportion of cases are related to surgical procedures and apparently represent infection of the operative site with toxin-producing staphylococcal strains. Consequently, surgical procedures must be remembered as possible predisposing conditions in asymptomatic *Staphylococcus aureus* carriers who acquire TSS.

SUMMARY

In this chapter three conditions that arise from the normal vaginal flora have been discussed. The organisms described are ordinarily able to establish a benign relationship with the host. Nevertheless, under certain and frequently ill-defined circumstances these organisms may cause local or extragenital symptoms. The clinician must appreciate the dual role of these organisms, recognizing that although the vaginal colonization is ordinarily inconsequential, the complications that arise because of these organisms may be significant. The main purpose of discussing these organisms and conditions is to establish the nature of the microbiologic and clinical relationships involved. Although some of these relationships do involve the pregnant or surgical patient, the subsequent chapters will take up those infectious diseases that are primarily either surgical or obstetric complications.

6

Surgical Infections in Gynecology

Sepsis Associated with Gynecologic Surgery

- Overview
- Source
- Microorganisms
- Clinical Features
- Diagnostic Considerations
- Predisposing Factors
- Complications

Infectious Complications in the Oncology Patient

- Overview
- Source
- Microorganisms
- Diagnostic Considerations
- Predisposing Factors

SEPSIS ASSOCIATED WITH GYNECOLOGIC SURGERY

Overview

The majority of gynecologic surgeries that carry a significant risk of infection are total or subtotal hysterectomies. Regardless of approach, either abdominal or vaginal, the surgical transection of the vaginal epithelium exposes the operative site to the microbial flora of the upper vagina. When combined with the local tissue trauma, presence of blood or serous fluid, and systemic depression of immune function as a result of the operative stress, the contamination may develop into an infection that may have a trivial or significant impact on the postoperative course. The need for postoperative antibiotic therapy, additional surgical procedures, further laboratory tests, additional time in the hospital, and added care by physicians and nurses make infectious complications of gynecologic surgery a major consideration in the human and financial cost of hysterectomy.

Today, of course, antimicrobial prophylaxis has reduced the rate of posthysterectomy sepsis to a fraction of the rate encountered before prophylaxis. At that time, about half of patients undergoing hysterectomy required antibiotics postoperatively. Not all instances of antibiotic use were for life-threatening infections, but this figure indicates that the infectious complications were at least of enough concern to warrant antibiotic prescription. In the first part of this chapter, the nature of postoperative infections in gynecologic patients will be discussed. The issue of prophylaxis, which figures prominently in gynecologic surgery, will be deferred until a later chapter to allow emphasis on aspects other than antibiotics. The point to be made is that antibiotics are the frosting on the cake of good surgical technique and good patient preoperative and postoperative care.

The principles regarding posthysterectomy infection apply to other gynecologic operations, although the risk of infection may be less with surgical procedures other than hysterectomy. Thus, in the case of conization of the cervix, endometrial biopsy or curettage, vulvar biopsy or surgical correction of pelvic relaxation, myomectomy, and laparoscopic procedures, infections may complicate the postoperative course. Because the focus of this monograph is microbiology rather than the operative procedures themselves, emphasis will be on the lessons learned regarding infection resulting from hysterectomy. It will be left to the reader to consider how the technical aspects of each operative procedure relate to the genesis of infections other than those that are sequelae of hysterectomy.

Source

Because the majority of infections related to gynecologic surgery arise from the normal flora of the lower genital tract, it is easy to overlook the importance of exogenous microorganisms. The source of exogenous bacteria will be described first before returning to the subject of endogenous contamination.

Patients may present for surgery already infected. Extant respiratory infections will be exacerbated, and compromised pulmonary function may be further altered to enhance susceptibility to infection. The urinary bladder harbors a significant number of bacteria in approximately 5-10% of asymptomatic women. Vaginal or cervical infection or chronic pelvic inflammatory disease may also be present in women who are candidates for hysterectomy. Before the routine use of antimicrobial prophylaxis, it was customary to postpone hysterectomy after conization of the cervix to allow the postconization infection to subside, thereby reducing the inoculum introduced into the colpotomy site. Dilatation and curettage in a patient infected with *Neisseria gonorrhoeae* may result in pelvic infection, and occult pelvic inflammatory disease may predispose a patient to infectious morbidity after elective hysterectomy.

Bowel flora may occasionally be involved in peritoneal infections. For example, if the bowel is nicked in pelviscopic surgery, peritonitis or intraabdominal abscess may ensue.

The majority of microbial contaminants present at the operative site are the organisms present in normal vaginal flora. Antimicrobial prophylaxis does not sterilize the vagina. As a consequence, the operative wound is contaminated by microorganisms in the patient receiving prophylactic antibiotics much as it is in the patient who has not received prophylaxis. Perhaps the most important aspect of the contamination of the colpotomy site and pelvic peritoneum is not the immediate exposure to bacteria, but the events that follow incision and contamination of the operative site. It is now known that in the days that follow surgery, a profound increase in many vaginal bacterial species occurs. The bacterial inoculum that elicits symptoms of infection may not be composed of the organisms that initially contaminate the wound. Rather, the inoculum may actually be composed of the much larger bacterial population that develops after the surgical procedure because of their ability to grow after being released from the vaginal epithelial surface to the deeper tissues. In any case, the ultimate source of the microbial contamination is the flora of the lower genital tract.

Microorganisms

The endogenous microorganisms that may be isolated from the infected sites in posthysterectomy patients include all of the organisms that are found in the normal vaginal flora. However, which organisms are responsible for the symptoms of infection is open to debate. Organisms that are considered to have minimal virulence, such as lactobacilli, diphtheroids, or alpha-hemolytic streptococci, may be present at the infected site. But usually other, more virulent species are present as well. It may be argued that the symptoms of infection are caused by one or more species of virulent organisms, and the less virulent organisms are innocent bystanders. Much remains to be learned about the interactions of microorganisms involved in mixed infections. Until a better understanding of microbial synergy is gained, it will be exceedingly difficult to incriminate one or a few organisms as etiologic in a mixed infection.

Some inferences have been made as to the most important organisms in endogenously acquired posthysterectomy infections. The most common gram-negative facultative organism is *Escherichia coli*. This is also the most commonly isolated enteric organism in the vaginal flora.

Much concern about *Bacteroides fragilis* has come from the experience of surgeons. However, the species of *Bacteroides* most often encountered by the gynecologic surgeon are *B. bivius*, *B. disiens*, and *B. melaninogenicus*. *B. fragilis* is involved in some posthysterectomy infections, but infection with this organism is more common in general surgery cases. *Bacteroides* species are noted for their ability to produce abscesses and have virulence attributes that are believed to account for their success as pathogens.

Another group of organisms attracting attention are the enterococci. These organisms are frequently found at infected sites, but their pathogenicity continues to be debated. Often, the role of certain microorganisms in mixed infections is deduced from the efficacy of antibiotic regimens used. The enterococci are generally resistant to cephalosporin-type antibiotics. On the basis of the success of therapy, which includes coverage for the enterococci, the conclusion has been drawn that the enterococci were contributing to the disease process.

Although soft-tissue infections after hysterectomy frequently are caused by the organisms mentioned above, these are certainly not the only species that are involved. Indeed, any organism may be involved in posthysterectomy infection. Occasionally, severe staphylococcal or streptococcal infections may occur postoperatively, and cases involving clostridial infections have been reported. Fortunately, organisms such as *Pseudomonas* and *Serratia*, which present therapeutic challenges in other medical and surgical patients, are uncommon in gynecologic infections primarily because they are uncommon as members of the normal flora.

Clinical Features

Infections may arise at any location that has undergone surgical trauma. By contiguous spread, the infectious process may extend beyond the site of primary contamination. Consequently, the locations and types of infections observed may be cataloged. In the case of laparotomy or abdominal hysterectomy, the skin wound may become infected either by endogenous genital microflora or by organisms from the skin or hospital environment. An infection involving the incisional site may also dissect along fascial planes and cause either fasciitis or an accumulation of pus within the tissue planes. Many of the infections occurring after hysterectomy are localized near the operative bed, as in the case of vaginal-cuff cellulitis. Contiguous extension of the contaminating microorganisms may result in pelvic peritonitis.

In addition to the soft-tissue infections that complicate hysterectomy or other gynecologic operations, abscesses of the involved tissues may develop. Generally, abscesses are considered to be late consequences of the host's attempt to contain extensive microbial contamination by walling off the organisms, effete white blood cells, and blood and serous fluids. Stitch abscesses usually involve skin organisms, whereas vaginal-cuff abscesses or abdominal abscesses and infected hematomas most often involve bacteria from the lower genital tract.

The concept that soft-tissue infections occur early in an infectious process and abscesses occur later was suggested by animal model studies that demonstrated that mixtures of anaerobes and aerobes implanted into the abdomen of rats resulted in an acute-peritonitis phase that was fatal in a portion of the cases. The dominant organisms in this phase were the gram-negative facultative organisms. Abscesses that had a significant anaerobic component tended to develop among the survivors. The utility of this model for understanding human infections is clear, although many authors have overinterpreted its significance.

In cases of posthysterectomy infection, the microbial contamination involves a different mix of organisms, including many gram-positive aerobic and anaerobic species. The primary site of contamination is the vaginal cuff, and the microbial load on the peritoneal surfaces is probably less than that on the cuff. The animal studies also used fecal material and barium to initiate the infectious process. Less foreign matter should be present in the operative bed of women who have undergone hysterectomy. The primary applicability of the observations from animal studies may be that early in an infection the faster-growing facultative gram-negative organisms may proliferate and cause significant symptoms of fever, pain, induration, and exudation at the operative site. Many of the septicemias that develop early will involve facultative organisms such as *Escherichia coli*.

Abscess formation may occur at the operative site or remotely from it. Regardless of location, however, it often only develops several days after surgery. Thus, it is not unusual to have a patient seek treatment for an intraabdominal abscess 5 or more days after surgery.

Diagnostic Considerations

Suspicion of postoperative infection must precede any attempt at diagnosis. A physician may be especially alert for signs of infection among patients who display factors that predispose them to infection, although the postoperative course of patients at lower risk will also be followed with due diligence. Monitoring the temperature chart, the condition of the surgical wound, and the overall condition of the patient generally provides the first indication of infectious complications.

Fever is one of the cardinal signs of infection and results from the interaction of the phagocytic defenses with the infectious agent. The leukocytes release interleukin-1, which is an endogenous pyrogen, and the interleukin-1 serves to raise the set point of the hypothalamic thermoregulatory center, causing fever. However, temperature elevation, particularly an isolated measurement, does not unambiguously determine the condition of the patient.

Normally a patient's temperature varies during the course of a day, and fevers during the first 24 hours after surgery may be related to the procedure. But the main concern over early fevers is still infection. One frequently reads in the literature that the definition of standard febrile morbidity is at least two temperatures of 38°C or higher observed at intervals at least 6 hours apart after the first 24 hours following surgery. But this definition is for statistical and epidemiologic evaluation and should not be construed to mean that early fevers are uniformly unrelated to infection. Indeed, early fevers may be indicative of very significant infections. Atelectasis may be responsible for postoperative fevers, especially after abdominal hysterectomy, and usually responds to good pulmonary care including deep breathing exercises. Early wound sepsis, exacerbation of a preexisting infection, sudden exposure to contamination, as in the case of damage to the bowel, or fulminant infection due to a highly virulent organism may result in florid symptoms early in the postoperative course.

Whether fever occurs early or late in the postoperative course, it should arouse suspicion of infection and attention should be given to the operative site. Evidence of infection or collections of pus at the abdominal wound or vaginal cuff or gross inflammation and induration should be noted. Needle aspiration may be used, especially at the skin wound to ascertain whether pus is present. The material obtained from these sites either by aspiration or swab should be Gram stained and cultured with due consideration for anaerobic organisms that may be present. The Gram stain is not always helpful, but because it is easily performed and may yield valuable information, it should be included as part of the fever workup.

Further evidence of infection should be noted at sites other than the operative site. Blood cultures should be taken if appropriate, and evidence of urinary or pulmonary infection should be sought.

Predisposing Factors

The factors that predispose the patient to infection are well known and will be noted here only insofar as they affect the success of endogenous microorganisms in causing infectious complications. Two categories of predisposing factors are noted: those related to the surgery and those related to the patient.

The technical aspects of vaginal and abdominal hysterectomy have a definite relationship to the development of infectious complications. The blood carries antibiotics, which may have been administered prophylactically before surgery, and also carries humoral host defense factors and the phagocytic cells that control contamination. Consequently, the adequacy of the blood supply is significant in preventing proliferation of the microorganisms present in the surgical wound. Blood supply is also important in maintaining the vitality of the tissues. Necrotic tissue becomes a superb growth medium for bacteria, and procedures that devitalize tissue tend to promote bacterial growth. Such surgical aspects include electrocautery, rough handling of tissues, crush injury to healthy tissue when placing hemostats, and strangulation of tissue when tying off more tissue than is necessary to control bleeding.

While emphasizing the importance of good blood supply, it is important to mention the negative effect of extravasated blood. Blood and serous fluid provide nutrients and iron, which enhance the virulence of bacteria. If the surgery creates a space that may fill with serous fluid, this area becomes a culture medium. For this reason good hemostasis is emphasized, and suction drainage of the vaginal cuff has been reported to reduce posthysterectomy infections.

Other aspects of the surgical procedure include the time of surgery, duration of anesthesia, and extensiveness of the procedure. The specific influences these aspects have on bacterial growth at the operative site are not clear, but they may be related to both local conditions in the wound and the systemic effects of the stress of the operative procedure on the immune system.

Wound infections of clean-contaminated wounds, the description most appropriately applied to abdominal hysterectomy incisions, have been studied extensively, and predisposing factors have been established. A preoperative antibacterial shower is beneficial, presumably because it reduces the number of superficial bacteria; but a shave preparation of the abdomen the night before surgery has an adverse effect on the infection rate. Microbiologically, this may be the result of the razor producing a low-grade infection of the skin, causing an increase in the bacterial inoculum of the wound. Adhesive drapes are associated with a higher wound infection rate than cloth drapes. It is known that the bacterial count on the skin is controlled by its moistness, and occlusive dressings will cause an increase in the skin flora. This fact may explain the increased wound infection rate associated with adhesive incise drapes.

Certain characteristics of the patient have been associated with higher infection rates, and for the most part a microbiologic explanation of these differences is not now possible. Some investigations have suggested that socioeconomic status is inversely related to risk of infection, although if this is true, the risk may result from other more physiologically based factors. Hysterectomy performed on postmenopausal women is associated with a lower infection rate than that observed among premenopausal women. There may be a slight difference in infection rate according to whether the hysterectomy was done in the follicular or the secretory phase of the menstrual cycle, with the latter associated with a lower propensity toward infection, according to some authors.

Patients who have any of a variety of illnesses, some of which may not be related to the need for hysterectomy, may be at risk for infection. Diabetic patients, patients with other endocrinologic disorders, and individuals with illnesses that result in compromise of the immune system are subject to an increased risk of infection. The need for prolonged care in the hospital before surgery may allow the patient to acquire a microbial flora from the hospital environment. Obese patients also are considered to be at greater risk of infection than nonobese individuals. As noted previously, preexisting infections represent another risk factor. Although a surgeon cannot choose only patients who do not display these risk factors, control can be exercised in that elective surgery may be postponed until a correctable problem is addressed (such as a urinary-tract infection or respiratory infection). If the risk factors remain at the time of surgery, the surgeon and nursing personnel may be attentive to any signs of infection during the patient's postoperative course.

Finally, although this chapter has avoided discussing antibiotic prophylaxis in order to draw the reader's attention to the host-microbe interactions in surgical infections, it must be stated that the demonstrated effectiveness of prophylactic antibiotics in reducing infection after vaginal hysterectomy has caused many surgeons to consider failure to use prophylaxis in women undergoing vaginal surgery as an unnecessary risk. The use of prophylaxis in abdominal hysterectomy is considered unnecessary by many surgeons, although this notion has recently been challenged and prophylaxis is being favored. Further details on antibiotic prophylaxis will be presented in chapter 14.

Complications

The soft-tissue infections that arise after surgery may not respond to therapy as expected because of septic pelvic thrombophlebitis, infected hematoma, or abscess formation. Ultrasonographic evaluation of pelvic masses may be of benefit, and adequate resolution of abscesses or hematomas will require drainage. If no abscess or hematoma is present and antimicrobial therapy that has adequate coverage for gram-negative facultative organisms and anaerobic bacteria is ineffective, the possibility of pelvic thrombophlebitis should be considered. This complication may be microbiologically related to the ability of some strains of *Bacteroides* to produce a heparinase. Adjunct heparin therapy may be required in such cases to achieve a satisfactory effect with antimicrobial drugs, but this therapy should only be instituted after appropriate testing of the patient's clotting profile.

INFECTIOUS COMPLICATIONS IN THE ONCOLOGY PATIENT

Overview

The patient with gynecologic malignancy, like patients with other types of cancer, is considered to be at high risk for developing infectious complications. Yet for a variety of reasons, the exact nature of the risk of bacterial infection in the gynecologic oncology patient has not been well defined. Perhaps the need to deal with the malignancy makes infectious complications seem less important. Nevertheless, the lesson from other specialties is that infection is the cause of death in about 40% of patients with solid tumors. Hematologic malignancy often has a higher mortality due to infection.

The emphasis of this chapter is primarily on the infectious complications of gynecologic surgery, although the infections faced by the gynecologic oncologist are related to both surgery and other causes. These include intrinsic problems that arise before therapy as well as infections that are a consequence of therapy.

Source

In the oncology patient, bacteria and other microorganisms reach the host tissues both from endogenous and exogenous sources. As we have recognized the unusual constellation of infectious agents that infect the immunocompromised patient who has acquired immune deficiency syndrome (AIDS), we are reminded that many organisms only cause problems in compromised hosts. These organisms certainly must be ubiquitous in the environment and the host, but they are of such low virulence as to be exceedingly rare causes of disease. Nevertheless, the immunocompromise experienced by the patient with gynecologic cancer is sufficiently different from that of the AIDS patient so that the organisms that cause her infectious complications are generally not the same as those that infect the AIDS patient.

The primary danger to the patient with a malignancy is from endogenous flora. The flora of the gut and lower genital tract may be disseminated at the time of surgery, as a result of tumor invasion, by contamination from a colostomy site, or possibly as the result of translocation.

The concept of bacterial translocation is relatively new and requires some comment. On the basis of concordance between microorganisms causing bacteremia and the gut flora of compromised individuals, the idea that bacteria may pass directly through the gut wall to infect remote sites was born. Animal studies have given some credence to this concept, but definitive studies that show that bacteria with unique markers appear at remote sites are needed. Translocation is believed to occur by the passage of viable organisms from the gut lumen to the mesenteric lymphatic and ultimately to the liver, spleen, or vasculature. The conditions putatively associated with bacterial translocation include altered ecology of the intestinal microflora; immunocompromise, including immunocompromise due to trauma or cancer chemotherapy; or a breach of the physical integrity of the gut.

The concept of translocation has been proposed as an explanation for burn sepsis and life-threatening infections in other compromised hosts, but has not been proved as the cause. If translocation does have a role in sepsis in cancer patients, selective antimicrobial suppression of the gut flora may be indicated. At present, however, the role of translocation is speculative, and the presence of peritonitis, bacteremia, or splenic or hepatic infection does not confirm translocation.

Infectious organisms may be acquired by a quasiendogenous mechanism as well. It is known that within the hospital environment, patients often undergo a shift in the skin and respiratory-tract flora. The acquired flora may not represent different species from those normally carried by the hospitalized patient. Rather, bacterial strains from the hospital environment, which may have antibiotic susceptibility patterns and virulence attributes that are different from those found in the community, displace the patient's normal flora. Endogenous infections may then arise from these newly acquired organisms.

The full range of exogenous organisms that cause problems within the hospital environment also threaten the patient undergoing cancer therapy. Some of these organisms are discussed in the subsequent section.

Microorganisms

It is unfortunate that the information identifying the microbial species involved in infectious complications of gynecologic malignancy is so limited. From the data available, it is apparent that the most frequently encountered organisms in oncology patients are the same as those encountered in patients without malignancy. After surgery, the oncology patient is infected by a larger range of bacterial species than are seen in patients with infection after gynecologic surgery for nonmalignant conditions.

Staphylococcal infections involving *Staphylococcus epidermidis* or *Staphylococcus aureus* are typical of skin wounds, and a large variety of other organisms, including anaerobic species, are found in skin wounds as well. Notably, in the various infectious complications of gynecologic malignancy, gram-negative aerobic or facultative rods such as *Pseudomonas aeruginosa*, *Escherichia coli*, *Enterobacter*, *Citrobacter*, *Klebsiella*, and *Proteus* are often isolated. The anaerobic isolates may include *Bacteroides*, the anaerobic cocci, and *Veillonella*. These organisms may be from enteric sources, and not uncommonly enterococci are also present. The possibility also exists that these organisms may be acquired in the hospital environment. This is of some importance because not only are organisms such as *Pseudomonas aeruginosa* resistant to many commonly used antimicrobial agents, but they may exist in the hospital environment, where selective pressures of frequent antibiotic use may increase their potential for antibiotic resistance.

Undoubtedly every surgical procedure is accompanied by a transient bacteremia, but in most individuals with intact defenses, these organisms are efficiently cleared from the bloodstream. Although direct comparisons are not available, it is probable that the transient bacteremias that accompany surgery result in frank septicemia more often in cancer patients than in other patients. Various gram-negative and gram-positive organisms including *Escherichia coli*, enterococci, *Klebsiella*, and staphylococci have been observed, in addition to other less common organisms. As noted already, in severely compromised hosts almost any organism may behave as a pathogen. Thus, one occasionally notes *Bacillus* sp. or *Saccharomyces cerevisiae* in infections of oncology patients.

Urinary tract infections are frequent complications in the patient who has undergone surgery. Although the vast majority of bacteremic women are infected with *Escherichia coli*, the organisms causing urinary tract infection in women with gynecologic malignancy are more diverse and include *Pseudomonas*, *Klebsiella*, enterococci, *Proteus*, *Morganella*, and possibly other organisms. These infections may be sequels of surgery or the result of indwelling bladder catheters. Urinary tract infection may involve both the upper and lower tracts. Because septicemia may arise as a sequel to urinary infections, appropriate diagnosis and therapy are essential. Diagnosis should include culture and susceptibility testing because one cannot presume that the offending organism is likely to be *Escherichia coli*.

Infections at or near the surgical site also complicate the postoperative course. Peritonitis, abscess, infected hematoma, and cellulitis involving lower genital or gut microorganisms are frequently polymicrobial and involve both aerobic and anaerobic species. The list of specific organisms is essentially the same as that for wound infections and septicemias. Again, the diversity of organisms that may be involved should be underscored. Intraabdominal infections may include various species of *Clostridium*, some of which may be histotoxic. Although clostridial infections do not always manifest themselves as the very aggressive clostridial myonecrosis, this eventuality should always be borne in mind. The very presence of clostridial species in a culture or the typical large, box-car-like gram-positive rods on the Gram-stained film of an infected site are not synonymous with gangrene, and the clinical picture should be considered in conjunction with the microbiologic picture.

Pneumonias may arise in postsurgical patients due to the compromise of the respiratory tract occasioned by intubation and pulmonary disability. The microorganisms that may cause pneumonitis include endogenous upper-respiratory organisms as well as exogenous organisms. In addition, gram-negative pneumonias, which include those caused by enteric organisms or nonfermentative, gram-negative organisms such as *Aeromonas* or *Pseudomonas*, occur with regularity and provide a challenging therapeutic problem.

Diagnostic Considerations

The patient with cancer, especially during therapy, is considered to be at risk for developing infection. Consequently, the physician should be on the alert for signs of infection. The signs of infection in cancer patients are the same as those in any other patients, although the intensity of these signs may be diminished in other patients. Perhaps more important is the possibility that signs of infection may be attributed to noninfectious causes. For example, pain associated with infection may easily be attributed to the patient's cancer. Fever may be drug or tumor related, but more likely is associated with infection. One study reported that only 2% of fevers in patients with pelvic tumors were related solely to the tumor and only 5% failed to defervesce with antibiotic therapy.

Severely neutropenic patients represent a special case. In one study neutrophil counts below 1,000/μL in individuals with gynecologic malignancy were associated with a 17% rate of fatal infection. Counts above 2,000/μL were associated with a 2% infection rate in a separate study. Thus, only severely neutropenic patients are at an inordinate risk. Another problem with low neutrophil count is that both fever, which is mediated by interleukin-1 release from white cells, and the white blood cell count itself are used clinically as indicators of infection. Practically, it may be somewhat more difficult to identify infection in a patient who is receiving radiotherapy or chemotherapy, because these are major causes of neutropenia.

Culture represents another appropriate means of diagnosis, and one that probably is underutilized in oncology patients. In addition to the genitourinary sites that ordinarily are of concern in the patient with gynecologic cancer, pulmonary infection should be considered, particularly if the patient has been intubated. Multiple sites of infection should not be overlooked in the evaluation. As much as possible, the sites cultured should not have a normal flora that confounds the interpretation of the results.

Predisposing Factors

As noted above, infectious complications in the oncology patient may be related either to the malignancy or to the treatment for the malignancy. A significant cause of fatal infections among individuals with genitourinary neoplasia is the tumor itself. Mechanical barriers established by intact anatomic structures and drainage of certain pathways are important parts of the host's defense. Cervical cancer may result in obstructive uropathy, and a small percentage of patients may have pyometra before therapy. Bowel obstruction occasionally results from ovarian cancer, and benign as well as malignant tumors may become infected as an extension of a pelvic inflammatory disease.

Tumor invasion of barriers may also result in infections in cancer patients before the institution of surgery, radiotherapy, or chemotherapy. Invasion of the bowel, for example, may lead to peritonitis or septicemia. Certainly, tumors differ in their aggressiveness, and a variety of confounding variables such as age of the patient and systemic immunosuppressive capabilities of the tumor need to be considered. The most recent survey available suggests that infection is a presenting diagnosis in fewer than 10% of cancer patients.

Damage to anatomic structures in the course of therapy is common and also predisposes to infectious complications. Radiation therapy has resulted in mucositis, adhesions and bowel necrosis, stenosis, or perforation. It should be added that radiotherapy has been refined since some of the earliest reports of complications and may not bear the same hazards as in the past. Reports from the last two decades have indicated that bowel reactions and cystitis from radiotherapy probably occur in fewer than 10% of cases.

Mucositis resulting from chemotherapy also predisposes patients to infection. Because the epithelia are cell-renewal systems, they may be somewhat more affected by cytotoxic drugs than tissues with a slower turnover rate. The normal flora of these tissues may become invasive as the mucosa becomes denuded.

The hospitalized cancer patient may have intravenous catheters placed for extended periods of time, and these catheters may predispose the patient to bacteremic episodes, as they do in patients with other types of illnesses. However, the oncology patient may be immunosuppressed either because of therapies or because the malignancy itself alters the immunologic responses of the host. Commonly, *Staphylococcus epidermidis* is incriminated in bacteremias secondary to intravenous catheters. Although this organism is considered to be of intrinsically low virulence, in the cancer patient it may be more threatening.

The implantation of intracavitary radium may disturb the integrity of the uterus and predispose the patient to infection caused by bacteria introduced at the time of the placement of the implant. The microorganisms may be derived from the patient's flora or may be from an exogenous source. An incident involving two patients who both acquired clostridial infections after the same contaminated probe was used to implant radium during successive operating-room sessions is described in the literature and illustrates the possibility of introducing contamination by this means.

Finally, surgical procedures are always sources of infectious morbidity. As noted in the earlier section of this chapter, a variety of technical aspects of surgical procedures contribute to the risk of infection. The duration of surgery, the length of time the patient is anesthetized, the amount of incidental tissue damage, the amount of suture material used, and the presence of dead spaces where blood and fluid may accumulate have all been identified as contributing to infections after surgery. Yet, as the tendency toward more radical operations increases, inherently greater risks of infectious complications will accompany oncologic surgery. Recent estimates have placed the rate of serious infectious morbidity in patients undergoing surgery for cancer at 15%. It should be remembered that in the literature one often reads that the prevalence of febrile morbidity for hysterectomy cases may be 20-40% or higher. However, this represents technical morbidity and may in many instances be trivial. Thus, a 15% rate of serious morbidity is significant. A higher rate of morbidity has been noted among patients with more advanced stages of cancer; although this is not surprising, the reason for this phenomenon has not been established.

A variety of surgical procedures uniquely associated with oncology practice present special risks for postoperative infection. Pelvic exenteration and radical dissections, including vulvectomy, cause significant tissue damage, with large areas of tissue exposed to potential microbial contamination. Anatomic perturbations also occur in the form of urinary conduit or colostomy. The former provides a portal of entry for microorganisms to reach the upper urinary tract, and the latter is a source of microorganisms that reach the skin at a site other than the perineum. The need to care for the colostomy also provides a source of fecal contamination to the patient or attendant.

7

Endogenous Obstetric Infections

In this chapter several infectious processes in the obstetric patient that primarily arise from endogenous flora are discussed. Not all occurrences of these infections are due to normal flora, and urinary tract infections are not limited to the obstetric patient. Because one of the major concerns of the specialty is urinary tract infection in pregnancy, it is included in this chapter.

URINARY TRACT INFECTION

Overview

The anatomy of the urethral meatus in the female is such that contamination from the vaginal or perianal flora may occur readily. Therefore, the most common urinary tract infection involves the lower tract. Cystitis may occur with symptoms of dysuria or may be asymptomatic. The latter condition is of primary concern in the pregnant patient because of its known association with prematurity. Less common but more dangerous is upper-tract infection, which usually represents a complication of lower-tract infection. Pyelonephritis is a serious condition at any time, but its presence during pregnancy is of special concern.

Source

Most uropathogens are believed to originate with the fecal flora, and as a consequence gram-negative facultative rods predominate in urinary-tract infections. However, recent evidence suggests that in women the vaginal introitus may be colonized with enteric organisms, and the presence of these organisms in the vaginal flora correlates reasonably well with the tendency to develop bacteriuria.

In addition to the characteristics of microbial colonization that can result in urinary infection, various perturbations of the urinary tract may result in infection. Placement of an indwelling transurethral catheter increases the likelihood that cystitis will occur. The longer the catheter remains in place, the more likely is the development of bacteriuria.

Microorganisms

Most urinary tract infections are due to *Escherichia coli*, although it is inappropriate to assume the etiology of infection in any patient. Uropathogenic *E. coli* are frequently characterized by the presence of mannose receptors, which are apparently involved in their adherence to the epithelium of the urethra and bladder. Because one of the protective mechanisms of the urinary tract is removal of organisms at the time of micturition, adherence to uroepithelium seems a logical virulence attribute.

Most surveys have identified *Klebsiella-Enterobacter* as the second most prevalent group of organisms responsible for urinary tract infection. Various species of *Proteus* rank third among the organisms causing uncomplicated urinary tract infection; however, infectious complication of obstructive uropathy frequently is attributed to species of *Proteus* or *Pseudomonas*. When urine culture demonstrates the presence of either of these bacterial genera, the possibility of an anatomic abnormality of the urinary tract should be entertained.

In a small percentage of patients with bacteriuria, staphylococci or enterococci are isolated. The latter may be associated with the use of cephalosporin antibiotics, which are not effective against these organisms. Frequently, the isolation of *Staphylococcus epidermidis* is assumed to be the result of skin contamination; however, when significant numbers are isolated or when the specimen is obtained by catheterization, the potential significance of the finding is increased. Relatively recently, group B beta-hemolytic streptococci have been discovered to be present in some urinary tract infections, particularly those in postpartum patients.

The microbial attributes that contribute to uropathogenicity are not well defined. The adhesive properties of *E. coli* have been noted, and other organisms may possess comparable adhesins. Many of the gram-negative rods typically isolated from the urine are motile, and *Proteus* is capable of utilizing urea. Whether these properties are directly related to the ability of these organisms to cause cystitis is not clear. Rarely is anaerobic culture performed on urine specimens, and most investigators do not believe that anaerobic bacteria play a significant role in urinary tract infections.

Clinical Manifestations

It should first be noted that significant bacteriuria may be present in some patients without symptoms. Asymptomatic bacteriuria occurs in about 5% of women of reproductive age, and the percentage tends to increase with increasing age. This condition is of particular concern in pregnancy both as an antecedent to upper-tract infection and because of its association with prematurity.

Symptomatic lower-tract infection is characterized by the typical signs of dysuria, frequency, and possibly hematuria and pyuria. When upper-tract involvement occurs, the patient may experience more severe symptoms including flank pain, fever, and elevated white blood cell count, in addition to lower-tract symptoms.

Predisposing Factors

Some of the factors that predispose patients to lower urinary tract infection have been identified, including the chronic placement of an indwelling catheter and surgical manipulations of the pelvic structures. Pregnancy predisposes approximately one third of the individuals with untreated lower-tract infection to upper-tract infection. During pregnancy, there is a physiologic hydronephrosis and a slower transit of urine from the kidney to the bladder, which may be involved in the propensity for lower-tract infection to develop into pyelonephritis.

Complications

The most serious complication of lower-tract infection is pyelonephritis. The risk of septicemia as a consequence of pyelonephritis is an ever-present consideration during pregnancy. Because of this serious possibility, the presence of urinary infection in pregnancy must be handled promptly.

An epidemiologic association between the presence of asymptomatic bacteriuria and low-birth-weight infants has been established by several investigations. However, the exact mediators that cause early delivery in women with bacteriuria have not been identified. Despite the many gaps in knowledge, it does seem clear that adequate screening for asymptomatic urinary-tract infection should be undertaken as a part of routine obstetric care, not only because of the association of bacteriuria with diminished birth weight, but also because about one third of patients with untreated bacteriuria will develop pyelonephritis. Noninvasive and relatively inexpensive methods are available for establishing the presence of asymptomatic bacteriuria, and relatively inexpensive and effective therapies can be offered. This makes urinary tract screening in pregnant women an appropriate standard of care.

Diagnostic Considerations

The simplest tests to indicate the possible presence of bacteriuria are the "dipstick" tests to detect urinary nitrite and leukocyte esterases. The former test detects primarily the reductive action of enteric gram-negative rods on the nitrate in urine to produce nitrite. This test is best done on first-voided specimens of the morning because bacteria in the bladder will presumably have had time to act on the nitrates. Unfortunately, not all bacteria will be detected by this test, so a negative test does not rule out significant bacteriuria. A further disadvantage is that such tests do not identify the bacteria causing the infection. As noted before, *Escherichia coli* is frequently involved in uncomplicated urinary tract infections, but other species may alert the physician to infection related to some abnormality of the urinary conduit system. The leukocyte esterase test indicates chemically the probable presence of white cells in the urine. However, white blood cells in the urine can be detected by microscopic examination and are not always indicative of urinary tract infection. The simple dipstick tests are better than no urinary screening at all, but they are inferior to culture techniques.

Direct microscopy is also useful in diagnosis of urinary tract infections. Bacteriuria of more than 100,000 organisms per milliliter is usually visible in unspun urine samples. Likewise, leukocytes may be observed, with 5-10 per high-power field being defined as pyuria, but such a finding alone does not unequivocally indicate infection.

Culture collection techniques usually include obtaining a "clean catch" midstream specimen. If the patient is properly instructed, an adequate specimen can be obtained in most instances. Bladder catheterization may be employed if necessary. Catheterization only carries a small risk of causing bladder infection if infection is not already present and so may be used readily in those cases when a patient may not be able to provide a good uncontaminated specimen. More invasive techniques such as transvaginal bladder puncture or suprapubic puncture are mentioned for completeness but are rarely justified.

Urine collected by the above methods should be processed promptly; if it is allowed to stand at room temperature, insignificant numbers of bacteria may proliferate and cause a false-positive culture result. If any delay in processing occurs, the urine should be refrigerated. The laboratory will quantitatively culture the specimen.

Quantitative culture is the rule for urine specimens, because a limited number of bacteria normally are washed from the skin and distal urethra in noncatheterized urine specimens. Consequently, quantitation is the method of establishing significance. The extensive studies of Kass indicated that 100,000 bacteria per milliliter of clean voided urine represent significant bacteriuria. The same level of bacterial colonization is significant in patients with symptoms of pyelonephritis. Recently, it has been established that colony counts of ≥100/ml are significant for specimens obtained by catheter or specimens from frankly dysuric women. As already noted, the bacterial species isolated is also important in assessing the significance of the culture findings. Although not always benign, *Staphylococcus epidermidis* may indicate skin contamination, whereas *Proteus*, *Serratia*, or *Pseudomonas* may indicate complicated urinary infection.

Within the hospital setting or large clinical office practice, the services of a microbiology laboratory may be readily available, but the obstetrician practicing in a small office setting does not have to forego urinary screening of pregnant patients. There are several easy-to-use, although not always easy-to-interpret, kits that allow quantitative culture to be performed without a fully equipped microbiology laboratory. The physician who utilizes such tests should recognize the limitations of the tests as well as the limitations of his own interpretation. It should be remembered that urine culture may be repeated either to verify the diagnosis or to test for cure. It is estimated that a single positive culture is about 80% accurate, whereas there is a greater than 90% certainty when two tests are positive.

CHORIOAMNIONITIS

Overview

Much confusion and controversy surround this condition, and in practical use the term "chorioamnionitis" includes more than inflammation of the chorioamniotic membrane. The inflammatory reaction may display limited involvement of the amniotic tissue or may involve the subamniotic zone of the placenta and the umbilical cord. Bacterial contamination of the amniotic fluid is also generally considered to be a part of this process, although it may be discussed by some authors under the separate terms of "intraamniotic infection" or "amniotic fluid infection." Chorioamnionitis is frequently associated with premature rupture of the fetal membranes and premature delivery. A disputed issue is whether these are independent events or causally related, and if they are causally related, whether chorioamnionitis is a cause or a result of premature rupture of the membranes and premature labor. Chorioamnionitis may be a histologic diagnosis that is not synonymous with bacterial infection of the membranes. However, this chapter is mainly concerned with bacterial chorioamnionitis and related infectious processes.

Source

There is little evidence to indicate that the bacteria that cause chorioamnionitis and related infections arise from sources other than the normal or abnormal vaginal flora. Although there is general agreement that chorioamnionitis is an ascending infection, a vigorous area of research and hypothesis is that altered vaginal flora may represent an especially significant source of bacteria that cause chorioamnionitis, amniotic fluid infection, and premature labor. An association has been described between patient-reported coitus near the time of delivery and an increased risk of preterm delivery and membrane inflammation. Although many limitations in these observations raise doubt about the reality of this association, it has been hypothesized that infection of the membranes is caused by microorganisms introduced at coitus, alterations in vaginal flora due to the effect of seminal fluid, or bacteria that are attached to sperm and carried upward. An association between bacterial vaginosis and prematurity has also been claimed on the basis of recent investigations. As noted in chapter 5, bacterial vaginosis is a condition of altered flora. Theoretically, this flora alteration exposes the membranes to greater numbers of specific bacteria or different bacteria than are present in women without bacterial vaginosis.

Microorganisms

Because chorioamnionitis is usually noted on histologic section rather than by bacteriologic confirmation, most bacteriologic studies have emphasized the culture of microorganisms from amniotic fluid. There are many technical problems in attempting to determine what microorganisms cause chorioamnionitis, although some techniques for culture involving sterilizing the surface and attempting to obtain cultures from the chorioamniotic interface have been employed. Microorganisms that reach the amniotic fluid usually do so by crossing the fetal membranes with or without membrane rupture. This process is probably accompanied by some degree of chorioamnionitis, and as a consequence amniotic fluid cultures may allow some inference to be made about the presence of chorioamnionitis.

In the case of the patient with ruptured membranes, one would normally expect that any of the bacterial species found in the lower tract might also be present in the amniotic fluid. More notable are the organisms that reach the amniotic fluid in the absence of membrane rupture. Older studies suggested that up to 10% of women might have minimal bacterial contamination very late in pregnancy. Specifically, when cultures of amniotic fluid from patients not in labor were evaluated, fewer than 10% showed contamination.

However, cultures taken from patients in labor who had intact membranes showed more than 10 bacteria per milliliter of amniotic fluid in more than half of the subjects tested. These organisms belonged to various anaerobic and aerobic genera and were similar to those present in the lower genital tract. The older literature indicates that *Escherichia coli* is the most common isolate from infected amniotic fluid, although in many places group B streptococci are more commonly identified. The strain of *Escherichia coli* that carries the K1 antigen appears to be a persistent colonizer when present and has the unfortunate propensity to cause neonatal meningitis. Likewise, the group B streptococcus appears to persist in a significant number of colonized women, presenting a risk of intraamniotic infection and neonatal sepsis.

In addition to the organisms that receive the most attention, careful microbiologic work aimed at identifying all organisms present in the amniotic fluid, rather than studies directed at ruling out the presence of one species such as group B streptococcus, indicate that a wide range of microbial species may be identified in the amniotic fluid of women with intact membranes. Attempts have been made to categorize microorganisms found in amniotic fluid as having either high or low virulence, and some degree of success in predicting outcome on this basis has been reported. However, it must be remembered that the fetus is in many ways more susceptible to the hazardous effects of microorganisms, and some that are considered to have little virulence, such as *Staphylococcus epidermidis*, cannot be dismissed as unimportant.

Mycoplasmas have been identified in amniotic fluid, and their significance as a cause of maternal or neonatal infection is not fully elucidated. Moreover, most laboratories do not routinely culture them. Consequently, culture of the amniotic fluid will not always provide the practical benefit the practicing physician desires.

One of the normal inhabitants of the lower genital tract, *Candida albicans*, has received little attention in the past as a cause of amniotic fluid infection. Recent information, however, suggests that it may be a more common cause of infection, especially in women with a suture placed for cervical incompetence. Cases of abortion, stillbirth, and neonatal death have been attributed to this organism.

Although rarely isolated from the female genital tract, *Haemophilus influenzae* has been reported to occasionally appear in amniotic fluid cultures and maternal and neonatal bacteremias, and it has been found in association with premature rupture of the membranes, spontaneous abortion, and amnionitis. This organism may continue to be isolated from these sources and should not be considered irrelevant. The virulence of the organism, as demonstrated by the experience of internists and pediatricians, suggests that obstetricians should be watchful for signs of increasing prevalence in the hospitals where they practice.

Clinical Manifestations

Clinical signs are of significance in diagnosing perinatal infections because of the lack of appropriate laboratory tests and because the presence of chorioamnionitis will influence the subsequent management of the patient, including decisions about the use of tocolytic agents, aggressiveness of management, use of antibiotics at cesarean delivery, and information given to the physicians responsible for the infant's care after delivery.

Premature rupture of the membranes is usually accompanied by some degree of chorioamnionitis and is probably in many cases preceded by chorioamnionitis. It goes without saying that the physician should become skillful in correctly diagnosing the presence of ruptured membranes. Objective tests for the presence of amniotic fluid in the vagina are of primary value. The obvious signs of foul-smelling amniotic fluid or purulence are probably absent more often than they are present.

Chorioamnionitis without rupture of the membranes is more difficult to discern before delivery. Such classical signs as maternal fever and uterine irritability are not uniformly present. Maternal leukocytosis is superimposed on the physiologic leukocytosis of pregnancy. Uterine tenderness is present in some cases, as is fetal tachycardia. None of these indicators is pathognomonic for chorioamnionitis and should lead the physician to suspicion and to any laboratory testing available. Histologic confirmation of chorioamnionitis may be informative in retrospect, but does not aid in the management of the patient. Finally, it must be noted that silent chorioamnionitis still presents all the risks of maternal and fetal complications that symptomatic infection does.

Complications

The major consequences of chorioamnionitis include premature delivery, perinatal sepsis, and postpartum maternal infection.

The manner in which chorioamnionitis is linked to prematurity remains controversial, and perhaps no single explanation will suffice to explain all cases of prematurity. The first problem lies in the association of chorioamnionitis, premature rupture of the fetal membranes, and amniotic fluid infection. Amniorrhexis was classically considered to be an antecedent to chorioamnionitis and amniotic fluid infection, and undoubtedly this sequence of events may occur. It is well known that amniotic fluid contamination increases over time after rupture of the membranes. However, the finding that bacteria may be cultured from the amniotic fluid of women with ostensibly intact membranes suggests that chorioamnion invasion may precede rupture of the membranes, and likewise rupture of the membranes is not required for commencement of labor.

In the past, danger to the fetus was considered to result from premature rupture of the membranes without immediate delivery, followed by fulminant intraamniotic infection that threatened the lives of both the infant and the mother. This sequence of events was termed the "amniotic infection syndrome," and it is still observed in certain population groups, including those in developing countries and in some inner-city hospitals that serve economically disadvantaged patients. Expeditious delivery is warranted in these cases. The bacterial contamination of the amniotic fluid is apparently substantial and infects the fetus by means of the external auditory canal, the oropharyngeal route, the skin, and the eyes. The result in the infant may be pneumonia, meningitis, septicemia, or other infections.

Premature rupture of the fetal membranes does not always result in such profound consequences. In some cases, it appears that rupture of the membranes is not accompanied by massive contamination of the amniotic fluid, perhaps because of host defense properties of the fluid. This may also indicate that histologic chorioamnionitis, which is very frequently present, is the antecedent of membrane rupture. In such a case, the infant is not in imminent danger, and conservative management is appropriate. Deciding which patient will develop fulminant sepsis, however, becomes one of the challenges of modern obstetrics (see "Diagnosis").

Labor and amniorrhexis may be separate issues. Certainly amniotomy may be used to induce labor, but labor may begin without membrane rupture. Investigators have attempted to determine what influences bacteria may have on membrane integrity as well as on initiation of labor. One theory is that the structural integrity of the membranes is affected by infection, possibly mediated by proteases excreted as bacterial exoenzymes. Enzymes from white blood cells that are recruited to the site of infection could also be incriminated as serving to weaken the chorioamniotic membrane.

The issue of the induction of labor has been viewed independently, and the prevailing theory is that bacterial enzymes produce phospholipases that cleave arachidonic acid from the structural phospholipids of the fetal membranes. The prostaglandin synthetase activity normally present is believed to convert the arachidonic acid to prostaglandins, which initiate labor. Another possibility is that the phagocytic cells may release prostaglandins. Thus, the bacteria that come in contact with the membranes have the theoretical ability both to weaken the membranes and to mediate uterine activity. Yet it must be emphasized that these are unproven notions as to the cause of prematurity, and appropriate controlled scientific investigation may uncover a more complex role for the microorganisms.

Whether vaginal microorganisms reach the amniotic fluid by crossing the intact fetal membranes or enter by means of a breach in the tissue, they are a threat to the fetus. Some organisms appear to be more virulent than others with respect to the fetus, and the concentration of these organisms in the amniotic fluid may likewise be important in the genesis of damage to the fetus. The range of neonatal outcomes may be from silent infection to stillborn infant. Good prognostic indicators are lacking, but some inferences may be made on the basis of quantitative cultures of the amniotic fluid, the reputation of the offending organisms (group B streptococcus being considered particularly dangerous, for example), the condition of the neonate at birth, and results of neonatal cultures. One should not take a narrow view of the virulence of the isolated organisms, however, because the neonate has limited host defense capability. Some organisms that are considered to be relatively innocuous in the adult may be a life-threatening hazard in the fetus.

Whether the fetus escapes damage or not, the presence of chorioamnionitis or amniotic fluid infection predisposes the mother to the development of postpartum sepsis. This is true whether she delivers vaginally or abdominally, although cesarean delivery will necessarily allow the infected amniotic fluid to contaminate the wound and the peritoneum. In a subsequent chapter, postpartum sepsis will be taken up in greater detail.

Diagnostic Considerations

After the suspicion of chorioamniotic or amniotic fluid infection is engendered by clinical observations (which, as noted above, are frequently not clear-cut), the frustration regarding the current state of objective diagnostic methods for these conditions becomes more apparent.

Because the amniotic fluid is in direct contact with the amnion and the fetus, and because the fluid is accessible either by means of transabdominal needle aspiration or by intrauterine pressure catheter, attention continues to be directed at finding some factor in amniotic fluid that will indicate chorioamnionitis and provide a prognostic indicator of the risk of severe maternal and fetal sepsis. Because amniotic infection may become fulminant, diagnostic procedures need to be rapid enough to influence patient management.

Leukocytic infiltration into the decidua capsularis, chorion laeve, and amnion and thus into the amniotic fluid is one of the characteristics that investigators have attempted to exploit. Early reports that one white blood cell per high-power field in amniotic fluid identified patients who would develop clinical chorioamnionitis proved less reliable in the hands of subsequent investigators, leaving doubt as to the value of this observation.

Gram stain of the amniotic fluid has also been considered as a means of rapid diagnosis of amniotic fluid infection. Although it is possible to find bacteria in the amniotic fluid even in instances in which neither the mother nor the infant develops sepsis, they are usually in very low concentration and are not visible on Gram staining of the amniotic fluid. Gram-positive organisms are generally relatively easy to identify on Gram stain. However, because all cellular and other debris is gram negative, gram-negative bacteria may be especially difficult to visualize, and these may represent as great a hazard to the fetus as gram-positive organisms. With respect to urine samples, observation has indicated that detecting bacteria by Gram stain is very unreliable unless counts are at least 100,000 organisms per ml of uncentrifuged urine. This figure is useful for urine because 100,000 organisms per ml represents the level of contamination that is considered significant. Quantitative culture of amniotic fluid, however, suggests that 100-1,000 organisms per ml have clinical significance in amniotic fluid, making Gram stain a theoretically insensitive method of evaluation. Despite these observations, many physicians use Gram stain effectively as an indicator of intrauterine infection.

The foregoing discussion should not be taken to indicate that direct observation of the amniotic fluid has no utility; rather, it indicates that there are significant limitations to this type of observation, and a negative test does not indicate that the patient is free of danger.

Another rapid, direct observation technique used for amniotic fluid is evaluation of the accumulation of bacterial metabolites, which may be detected by gas-liquid chromatography. In certain sterile body fluids such as cerebrospinal fluid, the presence of lactate has been used as an indication of infection, but in amniotic fluid it is not predictive. Other organic acids, including acetate, propionate, or butyrate concomitantly present with succinate, may be predictive of infection. However, results of other investigators have not been consistent, and because of the limited availability of chromatographic methods, this technique may not find much clinical utility.

Some investigators have employed urine dipstick tests designed to identify the presence of leukocyte esterase in testing the amniotic fluid of women suspected of having amnionitis. Some investigators have enthusiastically endorsed this test, whereas others have found it neither sensitive nor specific enough to be useful.

Although quantitative bacteriologic culture cannot be performed within a time frame appropriate for influencing patient management, it is the only method for identifying the nature of bacterial contamination in the uterus. As noted above, bacterial counts above 100 organisms per milliliter appear to have clinical relevance. The culture identification of bacterial contamination in the amniotic fluid may be analogous to blood culture. In the future, rapid, automated blood culture devices may prove to be of value in detecting contamination in the amniotic fluid. Any such use will have to give attention to the inactivation of endogenous antibacterial substances in the amniotic fluid.

Systemic host reactions to the presence of chorioamnionitis may also be helpful. One of the tests that has been correlated with chorioamnionitis is the test for elevated levels of C-reactive protein in maternal serum. Early reports of superb specificity of this test have not been uniformly verified. C-reactive protein is technically not specific for amniotic infection, but rather is an indicator of an inflammatory reaction. The appearance of specificity will depend first of all on its use in a highly selected group of patients, namely, those suspected of having amnionitis. As with any clinical test, it is also necessary to establish quality control and levels of sensitivity. At present, therefore, this test remains a research technique that does not have clear diagnostic significance.

Although several objective tests may be employed in patients suspected of having chorioamnionitis, none has proved definitive in either demonstrating the presence of infection or quantitating the risk to mother or fetus. For this reason a combination of objective observations and clinical evaluation will continue to be the diagnostic standard for some time to come.

POSTPARTUM ENDOMYOMETRITIS

Overview

Infectious complications of vaginal or abdominal delivery occur with regularity in clinical practice, although the rate of infection after cesarean delivery is 10-20 times higher than that after vaginal delivery. Overall, no more than 10% of deliveries are likely to be complicated by infection; however, in some centers the rate of infection after cesarean delivery may be well over 50%, particularly in teaching hospitals, large inner-city services, or tertiary-care facilities. In these hospitals cesarean delivery may performed in 20% or more of pregnancies.

A variety of infectious conditions may complicate the puerperium, and some of these actually have their genesis before delivery. For example, one of the predisposing factors to puerperal endomyometritis is chorioamnionitis. Not all postpartum infections are referable to the female pelvis, and the physician must recognize the possibility that febrile episodes may be related to pulmonary, urinary, breast, or abdominal wound infection. However, it may be expected that 90% of puerperal fevers are due to uterine infections, and in the present discussion attention will be focused on uterine infection.

Puerperal sepsis has a history as long and colorful as it is tragic. The institutionalization of childbirth was accompanied by the infectious complications of streptococcal infection from exogenous sources. It is now known that the group A streptococcus was carried on the hands of obstetric attendants and displayed unusual virulence, infecting the puerperal uteri of parturients and causing uterine and subsequently generalized sepsis and death. Changes in the obstetric practices, perhaps the intrinsic virulence of the group A streptococci, and undoubtedly other undefined factors have caused the situation to be drastically different today. Group A streptococci account for fewer than 5% of the occurrences of puerperal infection, whereas 15-30% of these infections are due to the group B streptococci. In addition, whereas the group A streptococci were primarily exogenous pathogens of profound virulence, the causative agents of postpartum sepsis today are largely endogenous organisms of lower virulence.

Source

As already stated, the organisms that infect the puerperal uterus are usually the same ones that constitute the vaginal flora. However, it should be remembered that one of the effects of vaginal delivery and operative procedures involving the female pelvis is a prompt and profound expansion in the vaginal flora (see chapter 3). This enlarged bacterial population may be the true source of most puerperal uterine infections. As will be noted below, postpartum infections may arise from the organisms that cause intrapartum infection. Although these organisms may ultimately arise from the vaginal flora, the development of chorioamnionitis or amniotic fluid infection may involve selection of a population of more virulent organisms that will then contaminate the puerperal uterus.

It is relatively uncommon for exogenous organisms to infect the puerperal uterus, although the possibility that virulent group A streptococci from exogenous sources might cause problems again in the future should not be forgotten. An outbreak in 1968 involving 20 mothers and 5 infants caused severe illness but no deaths, and extraordinary measures were required to control the outbreak. A similar occurrence with a different virulent organism is not beyond comprehension.

Microorganisms

As noted above, the majority of infections that involve the puerperal uterus or the adjacent tissues arise from the dissemination of the microbial flora of the lower genital tract. Consequently, the organisms commonly involved include gram-negative facultative rods such as *Escherichia coli* and the anaerobic rods, exemplified by various species of *Bacteroides*, primarily *B. bivius*, *B. disiens*, and *B. melaninogenicus*. It is common for several different organisms to be present simultaneously in the infected uterus, as indicated by cultures obtained by careful sampling techniques. In addition to the organisms identified above, *Gardnerella vaginalis* has also been identified and occasionally appears in blood culture from women with puerperal infection, suggesting invasiveness in some situations. Gram-positive aerobic cocci have also been reported frequently in cultures of the puerperal uterus, including the coagulase-positive and coagulase- negative staphylococci, group D streptococci, and various unnamed alpha-hemolytic streptococci. It is always difficult to ascribe significance to any particular species in mixed infections, and when an organism such as *Staphylococcus epidermidis* is isolated, there is always doubt as to whether it is a contaminant in the culture or a contributor to the infectious process. Synergistic interactions between organisms may allow microorganisms of low virulence to play relatively important roles in mixed infections. Anaerobic cocci including *Peptococcus* and *Peptostreptococcus* species have been cultured from the puerperal uterus, as has the gram-negative coccus *Veillonella*.

Clinical Manifestations

Fever may be the most common presenting feature of endomyometritis, although a decrease or cessation of lochial flow may occur up to 24 hours before the onset of fever. Pelvic examination should be performed on patients suspected of having postpartum infection and would reveal a tender uterus, although some uterine tenderness may normally be expected, and possibly a foul discharge. Signs of peritonitis may be present, including rebound tenderness and reduced bowel sounds. A soft-tissue mass detected early after delivery may represent a uterine hematoma. It is appropriate to consider the possibility that a retained placental fragment might be a source of the infectious process. The free drainage of the uterus is important, and if a tissue fragment has occluded the cervical os, it should be removed. If the infectious process is moving along a broad ligament, the uterus may be noted to be pushed toward the opposite side, whereas posterior spread results in a band of induration around the bowel and uterosacral ligaments that may be noted on rectal examination.

In the patient who is believed to have postpartum endomyometritis, it is essential to also watch for clinical signs of other complications such as septicemia and septic thromboembolism. Early therapeutic intervention may produce good results, but recrudescent illness in the form of pelvic abscess is possible. Treated patients should be instructed to seek attention if fever, pain, or other symptoms of infection occur after discharge from the hospital.

Predisposing Factors

One of the most important predisposing factors in postpartum infection is the presence of intrapartum infection. Consequently, the presence of chorioamnionitis, amniotic fluid infection, and prolonged fetal membrane rupture tends to dispose the patient toward the development of puerperal endometritis. The well-known aspects of surgery that predispose to infection include the effects of stress on the immune system; the presence of blood and serous material, which serve as bacterial growth medium; foreign body reaction to suture material; and the presence of devitalized tissue, which also serves as an anaerobic bacterial growth medium. These may be contributing factors that help to explain the relatively higher prevalence of endomyometritis in women who have had abdominal delivery compared with those who have had vaginal delivery.

Regardless of the mode of delivery, multiple vaginal examinations, particularly if membranes are ruptured, may add to the risk of infection. Internal fetal monitoring carries a theoretical risk of introducing lower-genital-tract organisms into the uterus. However, this risk appears only to be theoretical because evaluations of infection in monitored patients indicate that monitoring adds very little to the risk of infection, and many of the patients who require monitoring probably are already predisposed on demographic grounds to develop intrapartum or postpartum infection.

Complications

Even before significant signs of infection referable to the uterus are apparent, a patient may become bacteremic. The bacteremia may result from dissemination from a uterine site of infection, especially in women who have undergone cesarean delivery, or may be a complication of pyelonephritis. This septic complication may frequently involve such organisms as group B streptococci, enterococci, *Bacteroides* sp., *Escherichia coli* and other gram-negative facultative rods, and the anaerobic cocci. Mycoplasmas have also been implicated in postpartum and postabortion bacteremias, although the virulence of these organisms is not clear. The rate of bacteremia is relatively high (about 20%) among women with postpartum endometritis, although with proper therapy most otherwise-uncompromised individuals recover. There is a present danger in the postpartum patient that septicemia and septic shock may supervene. High temperature and leukocytosis (30,000/μL), prostration, disorientation, anxiety, unusually severe pain, and signs of hemoconcentration are evidence of developing septic shock. Appropriate aggressive action may include the surgical removal of the source of infection.

Septic pelvic thrombophlebitis is far less common today than in the past because prompt antibiotic therapy for uterine infection generally obviates this complication. However, it has been estimated that 1-2% of patients with postpartum infections may develop septic pelvic thrombophlebitis. Ovarian vein thrombosis occurring within the first 3 days after delivery is associated with acute pain and fever symptoms with a lower-right-quadrant mass. A high spiking fever with little abdominal pain and without pelvic mass occurring several days after delivery could also indicate a variant form of pelvic thrombosis. If this condition is inadequately treated, pulmonary embolization may also occur. The addition of heparin to the therapeutic regimen usually causes a significant improvement within 24 hours; if it does, heparin therapy is continued.

Ovarian abscess may also complicate spreading uterine infection. Because of the nature of this infection, the fallopian tube becomes involved from the outside in addition to the lumenal side. In contrast, the usual case with tuboovarian abscess that complicates pelvic inflammatory disease is infection from the lumen. Any abscess represents a threat because it may rupture and cause generalized sepsis, shock, and death. Spiking temperatures despite antibiotic therapy suggest an unresolving nidus of infection such as abscess. Various imaging techniques, including sonographic studies, computed tomography, or magnetic resonance imaging, may help identify the abscess. Drainage of the abscess is ordinarily necessary, although recent evidence indicates that abscesses of limited size may respond to medical treatment. Even when abscesses are treated surgically, antibiotic coverage is appropriate. The surgical techniques relevant to ovarian abscess treatment will not be discussed in this monograph.

Clearly, various serious and even life-threatening infections may complicate postpartum uterine infection. Occasionally, the complications are not adequately addressed by antimicrobial therapy, and surgical intervention in the form of laparotomy, laparoscopy, or hysterectomy may be lifesaving.

Diagnostic Considerations

Significant emphasis has been placed on the clinical observations in postpartum infections because it is primarily on the basis of symptoms that the physician suspects and begins to treat these infectious conditions. Because of the potentially disastrous consequences of postpartum infection, the patient who becomes febrile should be examined and a site of infection sought. Culture and susceptibility studies are controversial aspects of care for postpartum endomyometritis. It is not possible to obtain a truly uncontaminated uterine culture, because all culture devices must pass through the cervical os. Nevertheless, double lumen culture devices, which significantly reduce the contribution of cervical-vaginal flora to the specimens obtained, are available. Therapy should to be instituted on the basis of the most likely causative organisms, and the results of cultures may be used to make adjustments if therapy proves inadequate.

In cases where there is a possibility that the infection might involve clostridial species, a Gram stain is appropriate because these organisms have a distinctive gram-positive box-car shape. The presence of clostridia alone is not necessarily ominous unless accompanied by symptoms that suggest myonecrosis. Many other microorganisms commonly involved in postpartum uterine infection cannot be identified on the basis of microscopic morphology. Drainage from abscesses may be Gram stained to give a preliminary indication of the types of organisms present. Although one sometimes reads about the distinctive morphology of anaerobes, it should be noted that it is not really possible to distinguish between aerobic and anaerobic species on the basis of Gram stain.

OTHER POSTPARTUM INFECTIONS

Overview

It would be inappropriate to end this chapter without noting that infections that do not involve the puerperal uterus also complicate the postpartum course. In some cases the infectious agents are disseminated from the lower genital flora, but in other instances flora from other parts of the body or exogenous sources are encountered. The following paragraphs briefly describe some of these conditions.

Postabortion Sepsis

Uterine infection does occur in some women after spontaneous or induced abortion. In such cases infection may be endomyometritis due to the extension of the lower genital microflora, much in the same way infections develop after cesarean or vaginal delivery. However, if abortion occurs with retention of some of the products of conception, the disease may follow a particularly fulminant course.

A special situation occurs when products of conception are retained because this devitalized tissue provides a large amount of nutrient to enhance bacterial growth. This tissue also provides a low oxidation-reduction potential, which favors the growth of anaerobes. Inadequate attention to ensuring the proper evacuation of the uterus may occur when abortions are illegally performed by unqualified individuals or are performed under unsatisfactory conditions. Moreover, if the uterus is damaged by surgical instruments, it may become more susceptible to infection involving lower-genital-tract microorganisms. Under unsatisfactory conditions of illicit induced abortion, exogenous organisms may contaminate instruments that are inadequately sterilized. If a patient has had an illegal induced abortion, she may be particularly reticent to give an accurate history. Despite the availability of legal abortions in well-equipped clinics, the physician should bear in mind the possibility that a patient has had an abortion induced by any of a variety of inappropriate methods. If such is the case, the presence of organisms that are not usually found in the lower genital microflora, such as *Clostridium* sp., should be considered.

Pelvic Cellulitis

This complication usually occurs secondary to a cervical or lower uterine segment laceration. The lower genital microbial flora that contaminates the wound initiates a local lymphangitis (parametritis) in the tissues of the uterine corpus proximate to the injury. The perivascular lymphatics provide a conduit for the organism, which eventually causes a secondary peritonitis and may follow the ureter to the perinephric area or may follow the vascular supply to the buttock or thigh. Fascitis involving the thigh and buttocks and retropsoas infections, which can be very severe and debilitating, have been described in women who had pudendal block during vaginal delivery and may represent an infectious complication similar to those that arise from cervical lacerations.

Wound Infection

As with any surgical procedure, wound infection may complicate cesarean delivery. Causative organisms often include skin flora or genital flora. Wound sepsis may be manifested as wound cellulitis or wound abscess. Among the most common infectious agents are staphylococci or the mixed aerobic and anaerobic flora typical of the female genital tract. Other wounds that may become infected include vaginal and cervical lacerations, as well as episiotomy incisions. These wounds are more typically infected by the local flora of the lower genital tract. Wound infections have been discussed in the preceding chapter and the same principles apply, so post-cesarean-delivery wound infections will not be discussed in any further detail here.

Puerperal Mastitis

It is estimated that 2-3% of nursing mothers may experience breast infection, and that up to 5% of these infections may progress to breast abscess. It should be noted that fever during lactation has been attributed to breast engorgement in some cases. Maintenance of lactation is an important part of the prevention of breast infection, although infections nevertheless occur. The term "sporadic mastitis" is invoked in those cases in which the nipple is traumatized and microorganisms enter through fissures to cause cellulitis. In addition to local symptoms, sporadic mastitis may cause the systemic sequelae of pyrexia, tachycardia, malaise, headache, and anorexia. Sporadic mastitis usually occurs after the patient has left the hospital, and the physician may not be consulted. In contrast, epidemic mastitis usually arises early, even while the patient is in the hospital, and ordinarily occurs without prior trauma to the nipple. The invasive organisms enter the mammary duct system, resulting in adenitis with pus being expressed from the nipple and symptoms similar to those of mammary cellulitis. The course of this condition is more indolent than the course of mammary cellulitis and may be more protracted. The same strain of *Staphylococcus aureus* is usually responsible for individual outbreaks of the epidemic form, whereas various strains of *Staphylococcus aureus* are involved in sporadic cases. Other organisms less frequently encountered include streptococci belonging to groups B, D, F, and occasionally A, *Escherichia coli*, *Klebsiella pneumoniae*, *Serratia marcescens*, or *Haemophilus influenzae*.

When mastitis progresses to breast abscess, prompt evacuation of the collection should be undertaken, and the specimen should be cultured. Granulation and fibrosis may occur when abscesses are treated only with antibiotics, and breast deformity may result.

8

Perinatal Infections

The purpose of this chapter is to discuss several infections that arise in the newborn infant primarily as a result of exposure to the organism from a maternal source either while in utero or during the process of parturition. Characteristically, these infections are monoetiologic and are caused by organisms that are usually not considered to be normal flora. In addition, this group of organisms is notable because the injury that may be inflicted on the fetus or neonate tends to be disproportionately greater than the damage caused to the mother.

After recognition of the extraordinary fetal damage caused by the rubella virus, several other infections were noted to cause damage to the fetus as well. Most of these infections were acquired by the fetus through transplacental passage, although some can be acquired by other routes. The acronym "TORCH" thus was created to indicate that maternal acquisition of toxoplasmosis (TO), rubella (R), cytomegalovirus (C), or herpes virus (H) could result in fetal or neonatal damage or death. Of course, the term TORCH is incomplete, as reflected by the updated acronym "STORCH," which adds syphilis to this group of diseases and uses the letter "O" to indicate "other" organisms, some of which may remain to be discovered. The main emphasis of this monograph is on microorganisms other than viruses, and so the viral agents included under STORCH will not be addressed. Sexually transmitted diseases, which also may be included under the rubric of STORCH, are dealt with in chapter 9.

GROUP B STREPTOCOCCAL SEPSIS

Overview

When obstetrics was in its infancy, the infectious disease of greatest consequence was puerperal sepsis due to the group A streptococcus. In modern times the emphasis has shifted to the group B beta-hemolytic streptococcus, which usually has more serious consequences for the infant than for the mother. Maternal carriage of the organism may permit prenatal or perinatal acquisition of the organism by the offspring, with resultant generalized sepsis that may include pulmonary infection, septicemia, or meningitis.

The infection in the infant may follow a particularly fulminant course when acquired in utero and when symptoms develop in the neonate within the first 5 days after birth. These cases are described as early-onset infections and carry a very high rate of infant mortality, possibly as great as 50%, even when vigorous antibiotic therapy is given. Late-onset disease, which has a lower mortality rate, is nevertheless a serious complication that is associated with meningeal infection and in a significant number of cases permanent neurologic sequelae.

Source

In most instances, the group B streptococcus is acquired by the infant from the mother. The organism is present in the vaginal flora of up to 15% of women, although it may appear transiently. Thus, it is not possible to identify women who are carriers by a single vaginal culture because the organism may appear to come and go in vaginal cultures without apparent reason. Rectal carriage may be more stable and can serve to explain the intermittent nature of vaginal carriage but has not been demonstrated to predict neonatal sepsis.

Vaginal organisms are usually responsible for infecting the infant, although nosocomial transmission in the nursery occasionally occurs. It is probable that the routine attention to aseptic procedures that typifies modern hospital practice reduces the risk of nosocomial group B streptococcal infection. When the bacteria are in high concentration in the maternal flora and when the mother does not have antibody, the infant is at greater risk. Transmission may occur while the fetus is in utero by infection of the amniotic fluid, either related to ruptured membranes or by the apparent ability of this organism to cross intact membranes. Alternatively, the organism may be acquired by the infant through contact with the vaginal flora during parturition. The earlier in gestation the organism is acquired by the fetus, the more likely is the risk of serious disease. Thus, early-onset disease typically is more dangerous than late-onset disease.

Studies that investigated the epidemiology of group B streptococcal colonization also discovered that the male urethra can harbor the organism. However, what role, if any, male colonization plays in the persistence of female colonization during pregnancy is unknown. The ability of the organism to colonize the male may be important in attempts to use antibiotics to suppress vaginal colonization.

Microorganism

As all streptococci, the group B streptococcus is a gram-positive spherical organism that forms chains. This characteristic is of limited use for diagnosis in clinical settings. The ability to identify the organism in the flora of pregnant women is desirable, but Gram stain is not useful because other streptococci, including the anaerobic streptococci, are usually present and are indistinguishable from group B streptococci on Gram stain. One instance in which Gram stain may be useful is in examination of neonatal gastric aspirates. The gastric fluid of neonates does not contain a bacterial flora, and streptococci may be noted in infected infants.

Culture for the group B streptococcus usually involves primary isolation and then biochemical testing for presumptive grouping of the organism. A unique hemolytic reaction known as the CAMP test and hydrolysis of hippuric acid are the main features that permit the presumptive identification of a streptococcus as belonging to Lancefield's group B. Definitive evidence that the organism belongs to group B depends on the use of an immunospecific test. The normal methods used for cultural identification may require 24 hours for confirmation of the organism's identity. Consequently, various investigators have been working on methods for rapid identification of group B streptococci.

Selective culture media are now available that will allow the rapid growth, possibly within only 4 hours, of group B streptococci from specimens contaminated with normal flora. In addition, the use of specific antiserum for agglutination reactions allows rapid detection of the group B streptococci, although these tests are not considered routine at present and have not been validated and standardized for clinical diagnostic use.

Sophisticated studies on the antigenic structure of virulent group B streptococci have demonstrated five major antigenic types based on the capsular polysaccharide of the organism. Although serotypes Ia, Ib, Ic, and II are important causes of focal infections and septicemia in the neonate, type III appears to be the most virulent with respect to neonatal sepsis and meningitis. Many laboratories now include serotyping among the tests performed on clinical isolates of group B streptococci.

Clinical Features

In most instances group B streptococcal infection affects the neonate without concomitantly causing maternal symptoms, although, as noted in the preceding chapter, this organism may cause puerperal uterine infections in the mother. The fulminating life-threatening infections of the newborn, however, are the most significant of the clinical conditions caused by group B streptococci.

Neonatal sepsis follows one of two presentations described as early-onset disease and late-onset disease. Symptoms of the former arise within the first 5 days after delivery and have an average onset time of 20 hours postpartum. An infant that is unexpectedly flaccid, with low Apgar scores and rapidly developing signs of respiratory distress syndrome, should raise the suspicion of group B streptococcal disease. Septicemia, pneumonia, and meningitis are common in cases of early-onset disease, and as noted previously, mortality is quite high.

Late-onset disease appears after the first week of life and may occur in infants not colonized at birth. Meningitis is a typical manifestation of disease in these infants. The occurrence of this infection is not related to the predisposing conditions that are associated with early-onset disease.

Predisposing Factors

At the outset it should be noted that early-onset sepsis may occur in term infants who do not have any clearly apparent predisposing conditions. In addition, late-onset cases do not have the same predisposing conditions as have been identified for early-onset disease.

One of the most obvious conditions that predisposes to neonatal sepsis is maternal colonization. Yet maternal colonization is not the sole determinant of invasive disease. Only 3-15% of infants born to mothers who are culture-proven carriers of the streptococci have positive isolations of the organism from the external ear, nose, or skin after birth. A relatively small fraction of infants actually have invasive disease, although the disease rate varies among hospitals. Most surveys indicate that fewer than four infants have invasive disease for every 100 colonized mothers. Thus, other predisposing factors need to be identified.

Some correlation of neonatal susceptibility with maternal antibody has been identified. Immunoglobulin G (IgG) is able to cross the placenta, and antibody against type III group B streptococci also has been shown to cross the placenta, with corresponding levels in maternal and cord blood. The quantity of antibody needed to confer protection has not been established, and as a result no test is available for identifying the pregnancies that are at risk. It is important to recognize that superficial colonization of an epithelial site such as the vagina or gut mucosa may not be accompanied by a systemic humoral immune response. Thus, it is not surprising that fewer than 10% of women have serum antibody against all group B streptococcal serotypes, and only in the case of type Ia were colonized women more likely than noncolonized women to have serum antibody. For other serotypes, colonization did not appear to be associated with genesis of serum antibody.

Among the clinically identifiable risk factors are prematurity and prolonged rupture of the membranes. Obstetric problems and internal fetal monitoring by placement of a scalp electrode have also been associated with early-onset sepsis. Multiple gestations have been associated with an increased risk for group B streptococcal sepsis, inasmuch as the second twin has the same predisposing factors as the initially affected twin, including exposure to the organism. Thus, if one twin has symptoms of sepsis, a thorough evaluation of the second twin, including spinal fluid examination, is warranted.

Diagnostic Considerations

In complicated pregnancies or pregnancies with other predisposing factors, the obstetrician should be alert for signs of early-onset sepsis. As indicated above, there are no definitive tests to identify the pregnancy that will be complicated by group B streptococcal sepsis. Therefore, careful clinical observation and good pediatric support are essential for the well-being of the newborn. Vaginal cultures obtained during labor might be useful but ordinarily are only of use to provide a complete clinicopathologic picture and not a diagnosis.

Diagnosis on the basis of objective findings is problematic. The main emphasis is on methods that will identify women who are colonized near the time of delivery. It is known that antibiotic administration may protect the infant, but which pregnancies might benefit from antibiotic treatment and how to couple diagnostic and therapeutic methods are areas in which more work is needed. Further details about antibiotic use will be deferred until the chapters on therapy. The main refinements in culture techniques are aimed at identifying women in labor who are colonized. Methods that require growth of the organisms are still too time consuming to permit timely intervention, and cultures taken earlier in pregnancy have very limited value because of the intermittent nature of vaginal colonization. Direct antigen detection methods will probably become available for clinical use in the future and at least theoretically offer the greatest promise of clinical utility.

Beyond efforts to identify women who are carriers of group B streptococci at the time they begin labor, diagnostic efforts are mainly directed at the neonate. When the clinical situation suggests the need for an infection workup, examination of the gastric aspirate by Gram stain and culture of ear canal, umbilicus, and blood are performed. Evaluation of the cerebrospinal fluid also may be warranted. There are no distinguishing radiologic features that differentiate hyaline membrane disease from pulmonary streptococcal infection.

Prevention

Averting the exposure of the offspring to organisms that are potentially hazardous is one of the continuing challenges for obstetric researchers. However, this goal remains elusive with respect to the group B streptococcus. Antibiotic administration has been demonstrated to interfere with the vertical transmission of the organism, but because maternal colonization is often intermittent and infant exposure to the organism does not invariably result in disease, it becomes very difficult to know how to assign antibiotic therapy to women in labor. Various protocols are in use currently across the United States and elsewhere, but no absolute criteria may be offered at this time for use of such protocols. The foremost consideration for any physician contemplating the use of a preventive program is the magnitude of the problem in his or her practice. Giving antibiotics during labor in pregnancies that are deemed to be at high risk for development of sepsis is one possible approach. Because early-onset sepsis may in a significant proportion of cases not prove amenable to antibiotic therapy, such early intervention seems warranted.

The future will undoubtedly see attempts at developing vaccines that prevent group B streptococcal sepsis, but thus far an appropriate immunogen has not been discovered.

LISTERIOSIS

Overview

Listeria monocytogenes has been known since 1926 and is the cause of a bacteremic illness that is associated with an intense peripheral blood monocytosis and liver damage. The disease may be increasing in prevalence, or it may simply be more prevalent than is commonly believed. The most extensive experience with the disease has involved European investigators, but as North American physicians increase their awareness of the disease, listeriosis emerges as a definite part of the infectious disease picture.

Source

The reservoir of *Listeria monocytogenes* is domestic and feral animals. Ingestion of contaminated meat, milk, or vegetable products accounts for most occurrences of human infection. It is believed that in addition to overt disease, humans may experience either very mild symptoms or no symptoms with subsequent gastrointestinal carriage. Vaginal colonization is a matter that is open to debate, although seminal fluid samples from husbands of women who experience habitual abortion were culture positive in 3 of 60 cases. The infection of the fetus usually results from hematogenous dissemination of the organisms from the maternal bloodstream across the placenta. Ascending infection from the lower genital tract of the mother remains a remote possibility, or at least an uncommon occurrence.

Microorganism

Listeria monocytogenes is a gram-positive, short, rod-shaped organism that resembles diphtheroid bacilli. The staining tends to be irregular and Gram variability arises in older cultures. The organism grows adequately on laboratory media, but laboratory personnel need to be aware of its appearance and characteristics and should also be alerted by the physician who orders the culture that *Listeria monocytogenes* may be present. An extended period of time may be required to identify the organism. It appears to prefer room temperature to body temperature and both exhibits its typical "tumbling" motility and grows at the lower temperature. Its growth at 20°C is the basis for cold enrichment, which allows the organism in mixed culture to increase in numbers while other organisms that grow fastest at 37°C are inhibited. Cultivation on semisolid blood agar medium results in colonies that range from 0.2-1.5 mm and show beta-hemolysis.

Analysis of *Listeria* antigens has been employed for epidemiologic studies but does not play a role in the clinical management of the disease. Several strains of the organism have been identified based on somatic and flagellar antigens, and some of these antigens cross-react with other microorganisms.

Among the virulence factors identified for these organisms are a hemolysin known as listeriosis, a monocytosis-promoting factor (MPF), lipase, and NADase. Exactly what role each of these factors plays in colonization, invasion, and production of symptoms is not clear, although the monocytosis typical of this illness is attributed to MPF. The pathologic manifestations of the disease include granuloma formation and focal necrosis of various tissues. The role of monocytic white blood cells and the damage that may result from the expression of lipolytic enzymes need further clarification.

Clinical Features

Listeriosis affecting a pregnant woman may not be accompanied by serious illness. The patient may experience sudden chills, fever, malaise, generalized muscular pains, headache, and sore throat, all of which suggest an influenza-like illness. Although such a patient will be treated as a bacteremic individual, she may have a focal infection of the endocervix as well and therefore should promptly be given an appropriate examination. Occasionally flank pain and discolored urine may be observed, but costovertebral angle tenderness is absent. The patient may have diarrhea. The symptoms will usually abate spontaneously, but astute diagnosis and vigorous treatment are nevertheless required to avoid fetal complications. Abortion or stillbirth may result, or the infant may be born with disseminated listeriosis.

Neonatal sepsis is usually acquired by transplacental transmission and resembles infection with the group B streptococcus in that both early-onset and late-onset forms are possible. The infant infected in utero develops a bacteremic infection that results in miliary granulomatous lesions in many organs. These lesions are responsible for the older name of the disease, "granulomatosis infantisepticum." Skin lesions consisting of multiple small maculopapular or papulovesicular eruptions may be seen on the trunk and extremities. Microabscesses and scattered yellowish-white lesions are seen on the placenta.

Infants with early-onset disease usually present with respiratory distress. In addition to the respiratory signs of apnea, cyanosis, hyperthermia, and bradycardia, other organs may be affected, resulting in hepatosplenomegaly, conjunctivitis, and skin involvement. It has been suggested that systemic infection can result in excretion of the organism in the fetal urine, with subsequent contamination of the amniotic fluid; aspiration of the amniotic fluid by the fetus results in the organisms reaching the lungs and gastrointestinal tract.

Late-onset disease occurs in infants who are healthy at birth but who develop meningitis during the first few weeks of life. These infants are irritable, have difficulty in feeding, are febrile, and have bulging anterior fontanelles. The sequelae include mental retardation or hydrocephalus, although mortality may be as high as 70%.

Diagnostic Considerations

The most important aspect in diagnosing this disease when it is encountered during pregnancy is a high index of suspicion. The physician should not treat a woman with flu-like symptoms casually. The source of fever should be investigated, including blood culture, and the suspicion of listeriosis should be noted to the laboratory. It may be necessary for a physician to institute therapy without culture confirmation of this disease, because laboratory identification may require more time than that required for other organisms.

In the neonate, examination of the placenta (including histologic evaluation and cultures), gastric aspirate, and other sites as deemed appropriate should be done. The presence of gram-positive rods on Gram stain of the gastric aspirate may be useful. Examination of the spinal fluid may likewise be helpful, and Gram stain of centrifuged cerebrospinal fluid may be useful in making early critical decisions about management.

Predisposing Factors

Listeriosis has typically been a disease that is associated with immunocompromised hosts. Thus, exposure is not uniformly associated with disease. However, pregnancy seems to increase the susceptibility of women to infection with *Listeria monocytogenes*. With the number of immunocompromised individuals increasing as a result of organ transplantation, treatment of malignancies, and acquired immune deficiency syndrome (AIDS), one must bear in mind that diseases once considered to be rare are going to be seen more frequently. In addition, the role of animal reservoirs in the epidemiology of listeriosis should be remembered. The physician should assiduously evaluate the history of exposure to animals or animal products among women who are suspected of having listeriosis. Finally, the absence of predisposing factors or exposure to potential animal sources should not absolve the physician from the responsibility to completely and appropriately work up a pregnant woman with flu-like symptoms that could indicate listeriosis.

TOXOPLASMOSIS

Overview

In the United States parasitic infections often are neglected because they appear less commonly than in other countries. Toxoplasmosis probably occurs more often than it is recognized. The infection normally occurs without symptoms but causes significant sequelae in the fetus, although the damage may not be evident until the child is several years old. The infection is treatable, and neonatal sequelae are avoidable.

Source

Toxoplasma gondii is primarily found in animal reservoirs. Uncooked meat is a source of the organism in cultures that include it in their diet, although the organism does not survive freezing. A commonly cited source in the United States is domestic cats, which may acquire the organism by hunting wild rodents. Exposure of the gravida to aerosols from the cat's litter box can result in infection of susceptible individuals. The organism is ubiquitous, and in addition to wild rodents, animals that prey upon these rodents may become infected. Thus, contact with various animal species has the potential to cause infection in pregnant women. However, from an epidemiologic standpoint, animals other than domestic cats and dogs are rare sources of the organism.

Because the organism is fairly common in nature and because infection is accompanied by mild or no symptoms, a substantial percentage of the population has had immunologic experience with the organism without any knowledge of prior infection. In the United States, approximately one fourth to one third of pregnant women will have antibody to *Toxoplasma gondii*, although the majority of these do not represent newly acquired cases. Serologic surveys indicate that there is a gradual increase in exposure to the organism, as evidenced by greater antibody prevalence with increasing age. In North America by the fourth decade about one fourth of the population will show evidence of contact with *Toxoplasma gondii*.

Microorganism

Toxoplasma gondii is a protozoan pathogen of humans and animals that passes through three forms during its life cycle: the proliferative trophozoite stage, the tissue cyst, and the oocyst. The trophozoite may be stained with Wright's or Giemsa stain and is the form that serves as the antigen for serologic tests for toxoplasmosis. The trophozoite is an obligate intracellular inhabitant and invades virtually every type of mammalian cell. As the trophozoites multiply, they ultimately rupture the host cell and disperse to infect other cells.

During the process of multiplication, tissue cysts form. These structures persist as viable organisms throughout the life of the host. They appear to have a predilection for muscle and brain tissue, although they may be present in many other tissues as well. These forms appear to provide a means of long-term survival within the host, and they may infect another host that ingests the infected muscle tissue.

The oocyst is shed in the feces of members of the cat family 5 to 8 days after the animal becomes infected. Extremely large numbers of oocysts may be shed at this time, which explains the caution given to pregnant women regarding contact with excreta of cats, especially hunter cats that may acquire the disease from tissue cysts of wild animals.

Clinical Features

A newly acquired infection in a pregnant woman is accompanied by only vague symptoms. Lymphadenopathy occurs in about 10% of primary infections, but fever and fatigue are usually absent. The white blood cell count resembles that of mononucleosis, with the presence of atypical lymphocytes.

The disease in the neonate follows a complicated and insidious course. Although infection is initiated during intrauterine life, maternal infection does not invariably lead to fetal infection. When maternal infection is acquired early in pregnancy, the probability that the fetus will be infected is lower than if the primary infection is acquired by the mother in the third trimester. However, the severity of the disease is greater if toxoplasmosis is acquired early in the gestation, and stillbirth is more likely when the disease is acquired in the first half of gestation. An infant infected while in utero may be born without symptoms or may be born prematurely and with symmetrical growth retardation. The more significant sequelae include chorioretinitis, anemia, thrombocytopenia, hepatosplenomegaly, intracranial calcification, and microcephaly. As many as one half of infected babies will have subclinical disease, which may lead to a failure to recognize and treat the disease in order to prevent the later development of deafness, chorioretinitis, reduced intelligence, or epilepsy.

Diagnostic Considerations

The Sabin-Feldman dye test is used primarily to detect IgG antibodies in the serum of individuals who have been exposed to *Toxoplasma gondii*. Thus, a positive test indicates that a woman may be infected and may infect her offspring, but it does not distinguish between primary infection and residual antibody generated by a prior infection. A second test performed three to four weeks later will demonstrate rising titers in the case of primary infection. The patient whose titers from an earlier infection have stabilized at a relatively high level requires further evaluation. The indirect hemagglutination test and complement fixation test become positive later than the dye test, and titers obtained with these tests may still be rising when the Sabin-Feldman test is indicating stable titers. If these tests are not helpful, antitoxoplasma antibodies of the IgM class can be detected as an indication of recently acquired infection. The physician's state reference laboratory should be consulted to provide information on how such a test may be obtained.

HAEMOPHILUS INFLUENZAE

This organism is a short, pleomorphic, gram-positive rod that has as its distinguishing growth characteristic the need for X and V factors (nicotinic acid derivatives and hemin, respectively). *Haemophilus influenzae* is usually seen clinically as a cause of respiratory tract infection and meningitis in young children, but it rarely causes perinatal infection. However, several investigators recently have begun isolating the organism from maternal postpartum bacteremias and neonatal meningitis and septicemia. The conditions that predispose to neonatal sepsis seem to be the same as for other perinatal infectious agents, namely, premature rupture of the fetal membranes and chorioamnionitis. Spontaneous abortions have also been attributed to this organism.

The source of this organism is open to debate. *Haemophilus influenzae*, which is the most likely source of the organism, is occasionally found in vaginal cultures. However, other sources are possible. Although in the experience of most physicians this organism has not become a significant problem, in some areas 1-5% of neonatal bacteremias and 2-3% of maternal bacteremias are caused by this organism. In cases of sepsis in older children and adults, this organism presents a therapeutic challenge that most obstetricians and gynecologists have had the luxury of not experiencing. This organism may now be emerging as a new perinatal pathogen requiring both surveillance and attention to the therapeutic requirements of mothers and infants.

CANDIDA ALBICANS

Only brief mention of this organism will be made here because it is covered in detail in chapter 5. *Candida albicans* has been reported to be the cause of abortion, stillbirth, perinatal sepsis, and infant death secondary to chorioamnionitis. One apparent predisposing factor in the reported cases was the presence of a foreign body such as a cervical suture or retained intrauterine device. The source of the organisms is logically the vaginal flora.

ESCHERICHIA COLI

This common intestinal inhabitant has been described by some as the most frequent cause of chorioamnionitis and neonatal sepsis, although in many centers the group B streptococcus is more frequently identified. The presence of the K1 surface antigen is especially dangerous because it has a predilection for causing meningitis. Other strains of *Escherichia coli* may also cause neonatal sepsis and meningitis when acquired perinatally. In general, laboratories do not identify the O and K types of *Escherichia coli*. The reason why K1 types are associated with meningitis is unclear, but it may relate to the anticomplementary nature of the material comprising the antigen. The fact that *Escherichia coli* is normally found in the fecal and vaginal flora should not diminish the attention accorded the organism when it is isolated from the amniotic fluid, gastric aspirate samples, or other specimens obtained from the neonate.

MYCOBACTERIUM TUBERCULOSIS

The old problem of pelvic tuberculosis is described by classic texts and was relatively common before the disease was well controlled in the United States. Today, tuberculosis is mainly a problem among immigrants from Southeast Asia and the Caribbean. In addition, the atypical, rapidly growing mycobacteria are associated with disease in individuals who are immunocompromised as a result of AIDS. Therefore, mycobacterial infections may cause perinatal disease. Manifestations of the clinical disease in the offspring may range from acute sepsis and death to mild symptoms. Not all exposed infants will be affected, and further surveillance efforts are needed to firmly establish the nature of the disease acquired perinatally.

MYCOPLASMA AND *UREAPLASMA*

Many investigators believe that these bacteria, which lack cell walls and have very exacting growth requirements, have important implications for the outcome of pregnancy. However, the methods for studying these organisms have lagged behind those for studying more conventional organisms, leaving the precise role of *Mycoplasma* and *Ureaplasma* in dispute. As noted before, these organisms may be present in the vaginal flora of sexually active women, and prevalence increases with sexual activity and the number of partners. *Mycoplasma hominis* has been isolated from blood culture in a case of spontaneous abortion, from the lungs of spontaneously aborted fetuses, and from the amniotic fluid in cases of chorioamnionitis. Such associations appear to be uncommon, either because they are truly rare or because the necessary laboratory diagnostic facilities are not available or not used. Interest in the role of mycoplasmas as causes of or markers for women at risk of premature delivery is currently high. However, scientifically rigorous proof of such an association will be difficult to obtain, although epidemiologic associations are being drawn from some large survey studies.

Part III
Exogenous Infections

9

Sexually Transmitted Diseases

Gonorrhea
- Overview
- Microorganism
- Source
- Clinical Manifestations
- Complications
- Predisposing Factors
- Diagnostic Considerations

Syphilis
- Overview
- Microorganism
- Source
- Clinical Course
- Diagnostic Considerations

Chlamydial Infections
- Overview
- Microorganism
- Source
- Clinical Course
- Diagnostic Considerations

Minor Sexually Transmitted Diseases
- Chancroid
- Lymphogranuloma Venereum
- Granuloma Inguinale

Pelvic Inflammatory Disease
- Overview
- Microorganisms
- Diagnostic Considerations
- Predisposing Factors
- Complications
- Issues in Therapy

Before taking up the subject of specific disease entities, it is worthwhile to mention some of the common features of sexually transmitted diseases. The organisms that cause these diseases tend to be very successful as pathogens, becoming widely disseminated throughout the population but usually causing rather limited disease manifestations. An organism that rapidly causes the death of the host creates a problem for its own survival, whereas organisms that cause mild symptoms may have a longer association with the host and hence do not threaten their own survival by destroying and depleting the environment on which they depend. Moreover, the mildness of the symptoms allows the organisms to be transmitted more effectively to other hosts because the infected individual is not usually debilitated and may continue with normal activities.

Sexual transmission is an additional advantage to the organisms. Whereas their ability to survive outside the host is very limited, their transmission to uninfected individuals through sexual activity ensures that they will move from an environment to which they are well adapted into the same kind of environment. Most sexually transmitted pathogens have Homo sapiens as their only natural host, allowing them to be very specialized in terms of their virulence attributes. Humans have made various changes in their mode of living throughout history that have reduced the incidence and severity of various infectious diseases such as plague and cholera. However, it is almost certain that sexually transmitted disease will remain a part of the human legacy.

Although the primary manifestations of infection with sexually transmitted agents may be mild, several of these organisms can cause serious complications over the long term and may also be transmitted to the infants of infected mothers with devastating consequences.

Finally, although most sexually transmitted diseases are acquired by susceptible hosts through intercourse or other sexual activities, some infectious diseases may not always be sexually transmitted. Various types of vaginal infections may be sexually transmitted, but evidence that they are always sexually transmitted is lacking. It may be appropriate to use the term "sexually transmissible" to describe such conditions, but they are not the subject of this chapter, which focuses exclusively on those conditions for which the natural history is almost always acquisition through sexual contact.

GONORRHEA

Overview

This classical bacterial infection has been responsible for a worldwide epidemic during much of the last half of the twentieth century. About one million new cases of gonococcal infection are reported annually in the United States, and by most accounts this figure represents fewer than the actual number of cases occurring in this country. The infection may occur without symptoms or may cause mucosal infections at primary sites of inoculation. Various sequelae with serious implications are possible in both male and female patients and for the offspring of the infected gravida, although the latter does not involve transplacental transmission.

Microorganism

Neisseria gonorrhoeae are gram-negative cocci that have a kidney-bean shape and are usually seen in pairs with the flattened sides adjoining. These organisms are intracellular parasites and may be seen inside granulocytes obtained from Gram-stained smears of clinical specimens. The growth requirements for these organisms are relatively exacting. Growth on rich agar media, such as chocolate agar with vitamin supplements, is used for cultivation from clinical sources. *Neisseria gonorrhoeae* uses glucose by aerobic oxidation rather than fermentative pathways and also produces the enzymes cytochrome oxidase and catalase. Iron is required for growth and full virulence.

In recent years intensive investigations have provided insights into some of the molecular mechanisms of pathogenicity of these organisms. Among the first observations relative to virulence was the finding that the organism is capable of producing four colonial types on semisolid media. These were identified as T1, T2, T3, and T4 based on colonial size, flatness, and opacity. Types 1 and 2 possess pili and are fully virulent at mucosal sites. These colonial types, as expected, are the forms usually cultivated when the organisms are freshly isolated from infected individuals. The pili are composed of a single protein, pilin, which may undergo antigenic variation. The pili allow attachment to epithelial sites. Types 3 and 4 are nonpiliated and arise from subcultures of types 1 and 2 or from nongenital sites in humans.

Outer membrane proteins are also involved both in virulence and in determining the colonial type of the organism. Outer membrane protein PII is associated with the characteristic of opacity and also has a role in adherence of the cells to epithelium. Outer membrane proteins PI and PIII are components of the porin system of this organism. High- and low-molecular-weight variants of PI are known and are associated with symptomatic genital infections and disseminated infections, respectively. Antigenic variation among outer membrane proteins, like variation in pilin primary structure, leads to antigenic differences in the organism.

Bacterial products that apparently contribute to the pathogenicity of these organisms include immunoglobulin A1 (IgA1) protease and surface receptors that are able to derive iron from transferrin.

Advances in epidemiology have led investigators to examine certain nutritional requirements (auxotyping) as a means of categorizing different strains of *Neisseria gonorrhoeae*. Thus, it is known that those strains that have a propensity to become disseminated are auxotrophs for arginine, hypoxanthine, and uracil (A-,H-,U-).

Neisseria gonorrhoeae also may produce penicillinases, which are of concern for the future of therapy. Two types of resistance are known, including penicillinase-producing *Neisseria gonorrhoeae* and chromosomally mediated resistant *Neisseria gonorrhoeae*. The former are mostly plasmid-mediated producers of beta-lactamase, whereas the latter do not produce penicillinase, but rather alter the organism's penicillin-binding protein.

Source

Most primary gonococcal infections occur as a result of mucosal contact with an infected individual. Not only vaginal intercourse, but also oral sex and rectal intercourse, may expose a susceptible individual to the organisms, which attach to and penetrate the columnar epithelium. In women, the cervical epithelium is the primary site of infection; the vaginal epithelium is resistant because of its squamous character. In premenarcheal females the gonococcus may infect the vaginal epithelium and should signal suspicion of abuse. The source of the organism in cases of neonatal ophthalmia is contact during parturition with the organism harbored by the mother. Disseminated gonococcal infection or infection of the upper female genital tract results from a lower-tract infection in most cases.

A large amount of attention has been given in the past to the possibility that the organism may be transmitted by fomites. One occasionally reads of contrived experiments that permit the recovery of the organism from an inanimate object. However, in the end one essentially returns to the concept that the natural history of infection involves direct contact of infected tissue of one host with susceptible tissue of another host.

Clinical Manifestations

Exposure of the female genital tract to the gonococcus may have several outcomes. It is known that every exposure to the organism does not result in infection. The estimated number of organisms necessary to cause infection is about 100,000. If infection does occur, it may be asymptomatic or cause symptoms in 20-80% of infected women. Women may experience vaginal discharge, urinary symptoms including frequency and dysuria, fever, possibly abdominal pain, and in some cases proctitis.

The development of upper-tract infection as a consequence of lower genital colonization with *Neisseria gonorrhoeae* may lead to significant abdominal signs that require differentiation from other surgical diagnoses. Further discussion of upper-tract complications will be covered later in this chapter under the heading of Pelvic Inflammatory Disease. In approximately 1% of individuals, gonorrhea becomes disseminated and enters a bacteremic phase with symptoms of chills and fever. Arthritis involving the distal joints and tenosynovitis may result, and typical skin lesions are signs of disseminated gonococcal infection. The skin lesions, which usually are not numerous, involve the wrists, ankles, and elbows and start with a petechial lesion that becomes bullous and then develops a necrotic center. The bacteremia may be intermittent over a period of 3 days, and in the absence of treatment, the disease tends to settle in a single joint.

In perinatally infected infants, conjunctivitis that may lead to blindness is the most common type of infection. The potential seriousness of this disease emphasizes the necessity of the ophthalmic prophylaxis that is routinely practiced. It should be noted that the prophylactic regimens used are ineffective for established disease. In cases in which the gonococcus has contaminated the amniotic fluid and perhaps undergone some replication, the infant may be born with an infection for which silver nitrate prophylaxis is not adequate. Moreover, despite routine use of prophylaxis, gonococcal ophthalmia neonatorum occasionally arises. Therefore, the physician must not assume that this disease is a problem of the past.

Finally, other infectious complications that occur in perinatally infected babies should be noted. Mucosal contamination may result in systemic infection and a particularly destructive septic arthritis.

Complications

The most prominent sequela of gonococcal infection in women is upper-tract invasion. Pelvic inflammatory disease secondary to gonorrhea is common, and as is now known, may be accompanied by upper-tract infection involving organisms other than *Neisseria gonorrhoeae*. One episode of pelvic inflammatory disease predisposes a patient to future occurrences. In addition, the tubal damage that may result predisposes to infertility and ectopic pregnancy. Increases in the prevalence of ectopic pregnancy have paralleled the high rates of gonococcal disease and chlamydial disease in the U.S. population.

The ability of these sexually transmitted organisms to ascend from the primary site of infection apparently by contiguous spread across the endometrium results in several complications. The tubal inflammation, scarring, and stenosis are only parts of the problem. The process is now known to involve other lower-tract flora as well. As will be discussed further in the section on pelvic inflammatory disease, primary infection due to *Neisseria gonorrhoeae* or *Chlamydia trachomatis* predisposes to concomitant or subsequent upper-tract invasion by lower genital microflora. Intraabdominal infection involving the gonococcus may be manifested as peritonitis and in some women may cause perihepatic adhesions, which are eponymally described as the Fitz-Hugh-Curtis syndrome.

Predisposing Factors

Specific and nonspecific local immunity factors may reduce the susceptibility of some individuals to primary gonococcal infection. However, the organism shows variability in some of its surface antigens and an ability to evade some defense mechanisms, which permits individuals to become reinfected. Moreover, although anti-pilin antibodies produced by immunization seem to hold promise for future vaccination efforts, antigenic heterogeneity and the attainment of adequate mucosal antibody levels are problems that have frustrated attempts to develop a vaccine. Serum antibodies do not appear to furnish protection from mucosal infection. Consequently, the major predisposing factor in mucosal infection remains exposure.

Antibody and complement appear to be important factors in an individual's susceptibility to disseminated gonococcal disease. Certain strains of gonococci appear to be resistant to serum bactericidal activity, and some individuals who lack the terminal components of complement appear to be more susceptible to disseminated gonococcal infection.

Studies of the natural history of gonorrhea indicate that the probability of upper-tract invasion is increased when the primary infection is acquired near the time of menses. This suggests that loss of the endometrial tissue removes a host defense system that otherwise interferes with the ascending contiguous spread of the organisms from the endocervix to the fallopian tubes. Electron microscopy of the infective process shows that the organisms attach to the columnar epithelium and penetrate to the subepithelial layer. Unless the organism is a strain with a propensity to invade the bloodstream, spread beyond the subepithelial layer is unlikely. Consequently, the invasion of the upper tract occurs by contiguous spread rather than by a hematogenous route in most instances. The exact role of the endometrium in controlling ascending infection is not clear.

Diagnostic Considerations

Because gonococci have a distinctive morphologic appearance on Gram stain, including their presence within white blood cells, it might be concluded that Gram-stained smears of exudates would be useful as a diagnostic tool. In the case of male urethritis this is true, but in the female Gram-stained smears from sites with a substantial normal flora (including the endocervix, mouth, and rectum) are unreliable. Clinically valuable information may be obtained by Gram stain of gastric aspirates of neonates suspected of being infected in utero.

For the most part, culture remains the essential element for definitive diagnosis. Culture material should be plated directly on a suitable enriched selective agar such as Thayer-Martin. Because of the sensitivity of the organism to adverse environmental conditions, the culture should be placed into a carbon dioxide-enriched environment after plating. Several commercially available systems combine the necessary medium and the appropriate atmosphere, and both physician and clinical assistants should be familiar with the proper use of these culture materials.

In women, the endocervix should be cultured primarily, but other sites should not be overlooked. Because rectal colonization is also present in a significant proportion of women, rectal culture should also be done. For those physicians who want to avoid the cost of two separate cultures, both endocervical and rectal swabs may be inoculated onto the same plate. The laboratory must speciate the organism because meningococci and gonococci occasionally infect the same sites. In addition to enabling the physician to provide appropriate care for individual patients, susceptibility data are important in tracking the prevalence of penicillin resistance.

Because of the time required to diagnose gonococcal infection by culture-based methods, there is interest in rapid, antibody-based antigen detection methods. Solid-phase immunoassay procedures have been developed and are commercially available. These allow the detection of gonococcal antigens in patient samples within a few hours. However, these immunoassay procedures are not replacements for culture. They are not available in all clinical laboratories, and if they are used in a physician's office, questions regarding quality control must be raised. This issue will be debated not only with regard to testing for the gonococcus, but also with regard to other laboratory tests performed in physicians' offices by personnel who are not trained as laboratory technicians.

SYPHILIS

Overview

This sexually transmitted disease is caused by *Treponema pallidum* and is of particular concern during pregnancy because of the organism's ability to cross the placenta and infect the fetus. Developmental anomalies may result in infants infected in utero. The problem of syphilis in pregnant women, as in all infected individuals, is that it may go undetected and be characterized by prolonged asymptomatic periods in its natural history. The organism is not detected directly by growth on standard laboratory media, but is usually indicated by serologic testing.

Microorganism

Treponema pallidum is a spirochete that is morphologically distinct from the other human spirochetal pathogens such as *Borrelia* sp., which have coarse spirals, and *Leptospira* sp., which have very tightly coiled spirals and hooked ends. The treponemal body is so thin that it is not visible with conventional light microscopy. Typical chromatic stains do not permit visualization, although special silver-based stains can reveal the spirochetes in tissue specimens. More useful for direct observation is dark-field microscopy, which reveals not only the treponemal cell morphology but also its distinctive motility. Because most microscopes are not equipped with dark-field condensers, the physician should take the time to locate a dark-field-equipped microscope, which may be found in the hospital laboratory, a research laboratory, or a city or county health department.

The motility of the spirochete is an energy-requiring process in which metabolic energy is used to move the axial filaments that reside beneath the outer membrane. Although the axial filaments resemble flagella, they are not external appendages; rather, they lie alongside the length of the cell wall and are enclosed within the sheath of the outer membrane.

Because investigators have not been able to successfully cultivate these organisms on artificial media through many replicative cycles, studies of the physiology of *Treponema pallidum* have been extremely complicated. The organism was once believed to be a strict anaerobe; however, it has been shown to consume minute quantities of oxygen. It is known that the organism is exquisitely sensitive to penicillin, but because replication is slow, the bactericidal action of penicillin is also slower than that for conventional rapid-growing organisms.

Source

Primary infection with *Treponema pallidum* is almost always sexually transmitted, although the site of initial invasion may involve extragenital epithelial tissues. Contaminated blood also may occasionally produce primary infection. Because the disease is characterized by bacteremic phases, the source of infection for the fetus is transplacental passage of spirochetes from the maternal bloodstream. In the older literature it was asserted that transplacental infection did not occur before the fourth or fifth month of pregnancy. However, it has been demonstrated that the spirochetes may be present in abortuses of much earlier gestational ages.

Congenital syphilis is a twofold problem that includes acquisition of syphilis by pregnant women and lack of adequate prenatal care among these women. Rarely will women who receive prenatal care have signs of syphilis at the time of their clinic visit. Therefore, reliance on serologic screening is an essential element of good medical care for all pregnant patients. The majority of women who give birth to congenitally infected infants are from socioeconomically disadvantaged groups who typically lack access to prenatal care or who are not sufficiently cognizant of the need for prenatal care. Most women who obtain routine prenatal care are screened for syphilis as they should be, but errors or delays in follow-up occasionally occur, and in some instances women may have syphilis and not receive appropriate therapy. Furthermore, if the therapy given to the mother is inadequate, an infant may still become congenitally infected.

Clinical Course

Syphilis has been called "The Great Imitator" because its clinical manifestations are myriad and may resemble those of other illnesses. This, coupled with the fact that physicians today usually have less direct experience with the disease than physicians of the past, may make the disease less obvious on the basis of clinical observation alone. In addition, the infection is not always accompanied by symptoms. Therefore, whereas symptoms may arouse suspicion of disease, far greater reliance on serologic observations is the norm for modern practitioners.

The disease begins with a painless (unless secondarily infected) ulcerating lesion at the site of infection on an average of 3 weeks (range of 10-90 days) after contact with the organism. The erosion is described as a "hard chancre" because of its indurated margins.

In the female patient, primary genital lesions may be on the labia, but they may also be located intravaginally and consequently may go unnoticed. Regional lymphadenopathy with nonsuppurating, moveable lesions is also seen in patients with primary disease. If untreated, the primary lesions will heal in 3-6 weeks with a thin atrophic scar. Primary infection may occur at extragenital sites, including the anus, rectum, mouth, throat, tonsils, breasts, fingers, and around the eyes. These lesions may heal more slowly than genital lesions and may be painful.

At any stage the disease may become latent, a condition characterized by a positive serologic test for syphilis in the absence of symptoms and in the absence of cerebrospinal involvement. During the first year, if untreated or if spontaneous cure (which should never be assumed) does not occur, the disease may reappear as secondary syphilis. During the first 4 years of latency, the disease is considered communicable. In late latency, after 4 years, the immune response typically renders the disease unlikely to be communicated except by transplacental transmission to a fetus.

The secondary stage of syphilis usually appears 6-8 weeks after the primary lesion and is typified by mucocutaneous lesions. The disease is bacteremic and may involve almost any organ system, but it usually is manifested by lesions on the skin of the trunk and by hyperpigmented plantar and palmar lesions. Large, flat condylomata (condyloma lata) may appear at anogenital areas, and other mucosal sites may also have erythematous or gray-white erosions. Other skin manifestations may include a reversible loss of patches of hair.

Systemic manifestations of secondary syphilis include nonspecific symptoms of malaise, arthralgias, mild fever, headache, loss of appetite, anemia, and generalized nonpainful lymphadenopathy. Syphilitic hepatitis, meningeal irritation, and nephropathy are also possible. A complete description of all symptoms that have been reported for this disease is not within the scope of this chapter. However, it should be emphasized once more that these symptoms are valuable in arousing the physician's suspicion and prompting the use of definitive tests.

Late syphilis or tertiary syphilis occurs in about 15% of individuals who are not treated or inadequately treated for syphilis. This phase is characterized more by damage resulting from the immune reaction than by damage caused by the organism itself. Gummas are manifestations of late syphilis and include granulomatous complexes of the skin, joints, and possibly viscera or other internal organs. Cutaneous lesions may break down to form destructive, eroded lesions that heal with scarring. More serious involvement may include neurologic and cardiovascular sequelae. Recrudescences of secondary disease may occur several times in the first 4 years of the disease process, although relapses usually become less common as the length of time from primary infection increases.

Neurosyphilis has recently been found to be important in the pregnant patient. For reasons that are not understood, the likelihood of neurologic involvement appears to be heightened in pregnant women compared with nonpregnant individuals, and central nervous system involvement may occur more rapidly in pregnant women than in nonpregnant individuals. For this reason, evaluation for neurosyphilis should be undertaken in any pregnant woman who is diagnosed as having syphilis.

Congenital syphilis is the problem of greatest concern from the standpoint of the obstetrician, because of the severe consequences on infant development that can accompany transplacental transmission of spirochetes. The probability that the organism will be transmitted to the fetus is about 95%, and currently it is estimated that 1.5-1.8 cases of congenital syphilis will occur for every 100 cases of primary and secondary syphilis. In the United States each year, between 30,000 and 40,000 cases of primary or secondary syphilis will be reported.

Congenital syphilis is prenatally acquired when the spirochetes traverse the placenta. Because much of the damage caused by congenital syphilis appears to have an immunologic component, the transmission of infection later in pregnancy (that is, after the fifth month, when the fetal immune system is reasonably well developed) is associated with significant damage. Without treatment, one fourth of infected fetuses die before birth, and another one fourth die shortly after birth. The remainder of affected infants are said to have early congenital syphilis if symptoms appear within the first 2 years of life and late congenital syphilis if symptoms develop after 2 years.

Infants born with symptoms may have bullous or vesicular skin eruptions that may also involve the soles and palms. Rhinitis and hepatosplenomegaly are also indicative of early congenital syphilis. Condylomatous lesions around the anogenital region contain infectious treponemes. Numerous other multisystem stigmata are possible. Late congenital syphilis is also a multisystem condition that often involves extensive anomalous development of dentition and skeletal structures. Ocular, neurologic, and mucocutaneous evidence of syphilis is also possible, although in about 60% of affected infants the only sequel of congenital infection is a positive serologic test.

Diagnostic Considerations

The accurate diagnosis of primary or secondary syphilis depends on appropriate laboratory tests. Clinical signs are suggestive and should arouse suspicion, but the suspicion should be accompanied by appropriate serologic tests.

Upon acquiring syphilis, the host promptly produces anticardiolipin antibody, or reaginic antibody. It is not clear whether the antigen is composed of tissue components released as a result of damage done by the spirochete to the host or whether the cardiolipin antigen is a component of *Treponema pallidum*. The anticardiolipin antibodies elicited in patients with syphilis are primarily of the IgG or IgM classes and are not specific for *Treponema pallidum*. As a consequence, biologic false-positive reactions do occur, although a positive reaginic antibody test usually indicates that the patient has syphilis. Many of the conditions that cause false-positive reactions are infectious diseases other than syphilis or inflammatory conditions, including hepatitis, viral pneumonias, varicella, malaria, immunization, and even pregnancy.

The nonspecific tests are described as serologic tests for syphilis (STS) or nontreponemal tests and include the RPR (rapid plasma reagin) card test and the VDRL (venereal disease research laboratory) test. These have the advantage of being inexpensive and applicable to serologic screening programs. The VDRL has been adapted to quantitation, which not only permits the detection of anticardiolipin antibody but also allows some inference about the state of the disease. The VDRL (and other nontreponemal tests) become positive 1 to 3 weeks after the chancre appears, and titers diminish with effective therapy. Thus, the quantitative VDRL can be used to follow the effects of therapy.

Because the nontreponemal tests are not immunologically specific for *Treponema pallidum*, positive STS results must be followed by a specific treponemal test to ensure that the positive test was not a false-positive reaction. Several tests are routinely available that use *Treponema pallidum* antigens to detect the presence of specific antitreponemal antibody either through immunofluorescent or microhemagglutination methods. The physician should realize that these specific tests should be used to follow up any positive VDRL or RPR tests. Modifications of the FTA-Abs (fluorescent treponemal antibody absorption) test to distinguish between IgG and IgM antitreponemal antibodies have been developed. Because IgM does not cross the placenta, assay of specific cord blood IgM may be of value in identifying fetal infection. However, these sophisticated tests for antibody class are not generally available to most practitioners, and by the time the results are available from a reference laboratory, they will not have an impact on the care of the infant.

CHLAMYDIAL INFECTIONS

Overview

It is currently believed that *Chlamydia trachomatis* infection is the most prevalent of the bacterial diseases transmitted by sexual contact. *Chlamydia trachomatis* is biologically very different from *Neisseria gonorrhoeae*, but the natural histories of the diseases they cause are comparable. *Chlamydia trachomatis* is a mucosal pathogen, infecting primarily columnar epithelial cells. It causes primary endocervical infection; in addition, chlamydial infection may be the antecedent of upper-tract infection in the nongravid female and neonatal infection in babies born to infected mothers. Currently, the infection is not a reportable disease, and the sexual partners of individuals with chlamydial infection may be less efficiently treated than are those infected with gonorrhea or syphilis.

Microorganism

Chlamydia trachomatis is a bacterium with a cell envelope that resembles gram-negative bacteria, is inhibited by antibiotics, and synthesizes proteins and nucleic acids. However, it is an obligate intracellular parasite with a unique life cycle. The infective particle is the dense, metabolically inactive elementary body that attaches to the columnar epithelium and enters the host cell through a process of endocytosis. When inside the phagosome, the microorganism is transformed to become the reticulate body. This is the metabolically active form, although it parasitically uses the host's ATP supplies rather than synthesizing its own. The reticulate body undergoes several rounds of binary fission to produce 100-500 progeny within the phagosome. The infected cell is now characterized by a microscopically visible, iodine-stainable inclusion that contains the maturing reticulate bodies. The process of maturation involves the conversion of the progeny, which are in the form of reticulate bodies, to elementary bodies, which are released from the host cell by exocytosis. The released elementary bodies may then infect other cells.

The chlamydia have a thin cell wall that apparently has little muramic acid. Lipopolysaccharide is produced but may be chemically modified compared with that produced by more conventional bacteria. The lipopolysaccharide may contribute to the virulence of the organism and is part of antigenic display of the organism. Antigens that are genus specific (polysaccharide surface antigen) and species specific (two species of *Chlamydia* are known: *C. trachomatis* and *C. psittaci*) have been identified. In addition, serotypes subdivide *Chlamydia trachomatis* further. Within this genus are three types of lymphogranuloma venereum (LGV; see "Lymphogranuloma Venereum," this chapter): LGV-1, LGV-2, and LGV-3. There are 12 trachoma types. A, B, Ba, and C, which rarely cause genital infection, are more commonly associated with endemic trachoma; types D through K are isolated from genital infections and occasionally from eye infections.

Source

Primary urogenital infection in males leads to nongonococcal urethritis, which is transmissible to sexual partners and leads to the symptoms of cervicitis and urethritis. In these cases the source of the infectious agent is sexual contact with an infected carrier. Infants perinatally infected develop inclusion conjunctivitis or pneumonitis as a result of contact with the infected genital secretions during parturition.

Clinical Course

Upon exposure of an uninfected individual to an infected sexual partner, the chlamydia may attach to and invade the columnar or transitional epithelium of the cervix. The organism produces an inflammatory response that involves the superficial epithelium as well as the submucosa. The infection may be persistent and cause lymphocytic infiltration. The disease in women described as mucopurulent cervicitis may be indolent and may not be accompanied by significant symptoms. As with gonococcal infection, upper-tract involvement may follow, in which infection of the fallopian tubes occurs, and the disease process may extend to the pelvic peritoneum. Tubal inflammation with attendant fibrotic changes may lead to infertility or a predisposition to ectopic pregnancy.

When the disease coexists with pregnancy, several additional clinical conditions may result. First, there is a propensity for genital chlamydial infection to elicit premature birth. The infant exposed to the organism at the time of delivery may develop an exudative conjunctivitis involving one or both eyes 5-12 days postpartum. If untreated, the inclusion conjunctivitis may reoccur. The neonate may also develop otitis, mucopurulent rhinitis, or pneumonia. The infected mother may also develop postpartum fever, although it is difficult to unequivocally establish that chlamydia are responsible for the postpartum fever.

Diagnostic Considerations

Culture of the organism is the most definitive method of diagnosis. However, because the organism is an obligate intracellular parasite, it must be grown in cell culture. The usual medium is cycloheximide-treated McCoy cells. After the cell culture is infected with material from the patient, iodine-positive inclusions may be visualized or, alternatively, specific fluorescent antibody stain may be used to verify the presence of the organism. Some laboratories use a technique of blind passage in which the organisms are propagated through a series of cell cultures to increase their numbers, thus enhancing detection. Clearly, the laboratory methods required for cultivation of *Chlamydia trachomatis* demand a certain level of experience. Not all laboratories deem it important to devote the space, equipment, personnel, time, and money to educate personnel to the degree necessary to provide highly reliable chlamydia cultures. Thus, culture-based diagnosis becomes problematic for most physicians. Epidemiologic data indicate that millions of Americans become infected each year, but many physicians rely solely on clinical observation to diagnose and treat chlamydial disease. Unfortunately, a definitive diagnosis cannot be rendered on the basis of symptoms, and asymptomatic individuals will not be detected. For the physician who has suitable culture facilities, it is essential that the clinical specimens be collected in the provided transport medium precisely according to the specification of the laboratory.

Direct antigen detection methods are now available. These provide a means of quickly identifying individuals with chlamydial infection. An enzyme-linked immunosorbent assay adapted to microtiter plate technology is available for use in clinical laboratories. Also available are fluorescent antibody staining or solid-phase immunoassay methods that use monoclonal antibody technology. Fluorescent antibody tests require an experienced observer to provide optimum results, whereas the solid-phase immunoassay has been adapted for use as an office-based test.

The availability of rapid, direct antigen detection methods is not the ultimate answer to controlling chlamydial disease at present. The issue clinicians must address is who should be tested and by which method. No single guideline applicable to all practices may be offered. It is important to have a reasonably accurate approximation of the prevalence of chlamydia among the patients seen in a given practice. Knowing the prevalence will influence the aggressiveness of screening by individual clinicians.

MINOR SEXUALLY TRANSMITTED DISEASES

Chancroid

Haemophilus ducreyi is the causative agent of chancroid. Although there are only a few thousand cases of chancroid per year in the United States, the disease is far more prevalent in temperate climates and economically disadvantaged populations. It is also very likely that the disease is underrecognized and underreported. The bacterial pathogen may be cultivated with difficulty from clinical material on chocolate agar with vancomycin.

Exposure to the organism results in multiple, painful genital ulcers 2 days to 2 weeks after contact. This organism has a predilection for the regional lymphatics and causes suppurative inguinal node infection. These may spontaneously drain, but aspiration provides material for Gram stain, which shows gram-variable rods. Although it is considered to be a minor sexually transmitted disease, chancroid has the potential for causing greater problems and awareness of the condition is essential.

Lymphogranuloma Venereum

Although the disease-causing agent is a biotype of *Chlamydia trachomatis*, the disease is quite different from the more common type of chlamydial infection. Organisms of the LGV biotype are sexually transmitted and display a tropic effect toward the regional lymphatics. The major clinical manifestation is inguinal and femoral lymphatic buboes, which arise 1-4 weeks after primary invasion and are accompanied by inconspicuous genital lesions. Perirectal lymphatic drainage of the vagina leads to the complaint of proctitis in women. The disease is reported in only a few hundred individuals in the United States each year and is more prevalent in the tropics. The nodes in the groin may be somewhat painful and firm; later they become fluctuant and may be drained by needle aspiration. Untreated patients may have destruction of vulvar tissues or elephantiasis of the genitalia as a result of lymphatic stasis.

Granuloma Inguinale

This disease, caused by the encapsulated, gram-negative, nonmotile *Calymmatobacterium granulomatis* bacterium, is rarely seen in the United States. It apparently is not spread very readily and produces a chronic ulcerative disease of the skin and genitalia. It is possible that this organism is not always spread by sexual contact, as evidenced by the presence of lesions that are not in proximity to the genitalia. The ulcers that arise in the pubic area may be spread to other areas by autoinoculation and are susceptible to secondary bacterial infection. Wright or Giemsa stain of the infected material shows typical Donovan bodies, which are encapsulated bacteria in large histiocytic endothelial cells. The bacteria themselves display bipolar staining on Gram smears.

PELVIC INFLAMMATORY DISEASE

Overview

Pelvic inflammatory disease is included with the sexually transmitted diseases because gonococcal or chlamydial infection may be the antecedents of this condition. However, almost every aspect of pelvic inflammatory disease is clouded with controversy. Even the term "pelvic inflammatory disease" is disagreeable to many experts because it does not identify the specific location of the infective process. The term will continue to be used in this chapter unless salpingitis, endometritis, or ovarian infection is specifically meant. The very fact that the disease process was long inaccessible to direct observation and direct culture has been the source of controversy about pathogenesis and therapy.

As with other diseases already discussed, therapy will not be addressed here. However, the nature of the organisms causing the condition will have a great bearing on the rationale for selecting therapy.

Our current understanding of pelvic inflammatory disease is largely based on laparoscopic observation, which has opened new dimensions to understanding pathogenesis and therapy. Yet there remain gaps in our knowledge of this condition. Perhaps the most important point to emphasize is the diversity of conditions that are included under the broad term "pelvic inflammatory disease."

It is almost trite to recount the fact that pelvic inflammatory disease is one of the most common medical problems faced by reproductive-aged women today and that the cost of medical care, including hospital expenses and surgical procedures as well as lost work time, is immense. The emotional stress of infertility, dyspareunia, and other human consequences can hardly be reckoned with. This very simply is one of the most important disease entities that must be treated by the gynecologist.

Microorganisms

It has only been since investigators have used laparoscopy for intraabdominal specimen collection that a reasonably complete view of the microbiology of pelvic inflammatory disease has emerged. The earliest view of the disease was that the abdominal symptoms were primarily the result of an infection of the fallopian tubes by *Neisseria gonorrhoeae*. The finding that *Chlamydia trachomatis* was a significant pathogen and could be an antecedent to pelvic infection opened a new chapter in the understanding of pelvic inflammatory disease. Currently, cultures taken intraabdominally via the laparoscope reveal that the gonococcus and chlamydia are inconsistently recovered, whereas organisms that seem to have originated in the lower genital tract are commonly isolated. *Mycoplasma hominis* and *Ureaplasma urealyticum* have been reported in intraabdominal cultures, as have *Gardnerella vaginalis*, *Escherichia coli*, group B beta-hemolytic streptococci, and nonhemolytic streptococci. Anaerobic bacteria are also present in a significant proportion of cases, and these organisms include *Bacteroides bivius*, other species of *Bacteroides*, *Peptococcus asacharolyticus*, and *Peptostreptococcus anaerobius*. It is important to note the presence of anaerobic bacteria not only because it is critical in planning antimicrobial therapy but also because it suggests why tuboovarian abscess commonly complicates pelvic inflammatory disease. The abscessogenic qualities of certain anaerobic bacteria was noted previously in chapter 3.

One of the unsolved questions regarding the polymicrobial nature of pelvic infection is whether the sexually transmitted organisms cause damage to structures of the upper genital tract (particularly the fallopian tubes), which then become susceptible to secondary invasion by other organisms from the lower-genital-tract flora, or whether both normal flora and sexually transmitted pathogens reach the upper tract simultaneously and cause disease. Damage to tubal epithelium has been attributed to *Neisseria gonorrhoeae* as a result of observations on explanted tubal samples. It is easy to understand how such damage could prepare the tissue for secondary invasion by organisms from the normal flora. Furthermore, it may be argued that because organisms from the normal flora do not cause pelvic inflammatory disease in the absence of an inciting sexually transmitted infection, intraabdominal infection is a two-stage process. Despite this, studies with *Bacteroides* in human fallopian tube organ culture reveal it to be capable of causing significant destruction of tubal epithelium. Nevertheless, definitive experiments establishing the natural history of monoetiologic or polymicrobial pelvic infection remain to be completed.

Diagnostic Considerations

Clinical observation is generally the most significant first step in diagnosis of pelvic inflammatory disease. Numerous studies have tabulated the symptoms present in patients with acute pelvic inflammatory disease, and some of these studies have correlated clinical findings with laparoscopic observations. The accuracy of diagnosing pelvic inflammatory disease on clinical grounds alone is probably less than 75%. Conversely, laparoscopy of patients not suspected of having pelvic inflammatory disease revealed that some, in fact, had salpingitis. In an attempt to provide some common basis for clinical diagnosis, the characteristics identified in Table 9-1 may be used along with knowledge of the factors that predispose to pelvic inflammatory disease to identify women who are likely to have the disease. Other factors have been used with varying degrees of accuracy in diagnosis of pelvic inflammatory disease, including erythrocyte sedimentation rate and C-reactive protein.

Definitive identification of the nature and extent of disease involves laparoscopic examination, which not only allows the diagnosis to be verified but also permits grading of the condition. Table 9-2 summarizes the laparoscopic findings and their interpretation. Obviously, because of economic and other considerations, laparoscopy will not be used in all cases. Therefore, the physician must bring clinical acumen to bear on the diagnosis

Table 9-1. *Criteria for clinical diagnosis of salpingitis*

Patient Must Have All of the Following:	Plus at Least One of the Following:
• Abdominal direct tenderness with or without rebound tenderness • Tenderness with motion of cervix or uterus • Adnexal tenderness	• Temperature >38°C • White blood cell count of >10,000 • Purulent material in abdomen by culdocentesis or laparoscopy • Gram stain of the cervix consistent with gonococci • Pelvic abscess or inflammatory complex demonstrated by sonography or bimanual examination

Table 9-2. *Laparoscopic Grading of Pelvic Inflammatory Disease*

Grade	Findings
Mild	Tubal erythema, edema, no pus expressed with tubal manipulation
Moderate	Gross pus evident, greater degree of edema, tubes may be immobilized, fimbrial stoma possibly nonpatent
Severe	Pyosalpinx or inflammatory complex, abscess

in all cases and laparoscopic observation in some cases, particularly those cases in which the diagnosis is in doubt.

Predisposing Factors

Acute pelvic inflammatory disease is a condition affecting sexually active women. Those who have had a previous episode of pelvic infection are at higher risk for developing subsequent infection. In addition, data on fertility suggest that multiple episodes of pelvic inflammatory disease seem to have an additive effect with regard to tubal damage. It is possible that a previous infection with *Neisseria gonorrhoeae* may also predispose to salpingitis because the organism may have caused subclinical salpingitis during the first encounter. Wearing an intrauterine contraceptive device has been identified as a risk factor, and the specific type of device has been correlated with propensity to develop infection. The type of device may be of diminishing interest in the United States because of the dwindling number of available devices. However, elsewhere in the world the issue is probably still relevant. In contrast to intrauterine devices, barrier contraceptive devices and the use of oral contraceptives have been associated with a lower risk of pelvic infection compared with that observed in women who used no contraception. Another risk factor that has been shown to influence the likelihood of developing pelvic inflammatory disease is young age (most cases are in patients aged 16-24 years); if corrections are made for sexual experience, the age group at greatest risk are sexually active individuals under the age of 15 years. Multiple sexual partners and sex with males who have untreated infection are also risk factors. Habitation patterns are associated with risk, with single women who live alone being at greatest risk.

Complications

As noted previously, pelvic inflammatory disease is not synonymous with salpingitis, and the various degrees of involvement result in a variety of consequences. Salpingitis is accompanied by differing degrees of tubal damage that may or may not be reversible. The scarring and stenosis that may follow an episode of salpingitis may result in nonpatency of the fallopian tube, leading to involuntary infertility. The tubal pathology may also predispose to ectopic pregnancy, and as stated before, the pathologic consequences of multiple episodes of tubal infection appear to be cumulative.

During acute pelvic inflammatory disease, the infectious process may extend beyond the tubal structures and cause peritonitis with subsequent adhesion formation. Tuboovarian abscess also is a relatively common consequence of pelvic inflammatory disease. It has been observed that 15% of women with symptoms sufficiently severe to warrant hospitalization for acute pelvic inflammatory disease developed tuboovarian abscess.

Tuboovarian abscess has typically been a surgical problem, but current observation has modified thinking about how to manage this condition. Surgical approaches have become increasingly conservative over the years. Whereas complete hysterectomy and salpingo-oophorectomy were frequently the standard of care, surgeons have moved toward unilateral oophorectomy and more recently to early aggressive medical management with antibiotics effective against the organisms found in such abscesses. Success has been reported in cases in which the abscesses were not large. Newer methods of abscess drainage include percutaneous aspiration and aspiration through the laparoscope.

The most dangerous of the complications associated with pelvic inflammatory disease is ruptured tuboovarian abscess. Aggressive treatment is essential under such circumstances, although surgical procedures have become somewhat more conservative over the years. With adequate antibiotic coverage and appropriate surgical treatment of the infected material, acceptable outcomes have been reported.

Issues in Therapy

Medical therapy for pelvic inflammatory disease will be discussed later. However, because pelvic inflammatory disease is heterogeneous in terms of etiologic organisms, the mixture of organisms involved, the various tissues that may be affected, and the severity of the disease at the various infected sites, it is very difficult to scientifically compare the response of one patient to therapy with the response of another patient. It is difficult to compare the effectiveness of one therapeutic regimen with that of another. Thus, questions of which therapy is best, how long therapy must be continued, and how therapy should be tailored to the mixture of organisms present in individual patients present challenges for the future. Attempts have been made to score the severity of symptoms to provide some quantitative basis for comparison of therapies; however, it remains to be seen whether such scoring systems are artificial or true reflections of the efficacy of therapy.

Currently, physicians are faced with a standard of care that directly conflicts with attempts to enforce cost containment. Good care requires a willingness to hospitalize and laparoscope a significant number of patients in order to provide aggressive treatment that will preserve fertility and prevent the need for more extensive surgery later. Although emphasis is placed on outpatient care as much as possible, some patients, especially those who are very young, will need hospitalization to permit close observation and to ensure complete compliance with antibiotic therapy. These measures will preserve fertility and prevent the progression to more serious stages of disease. Judicious use of hospitalization, laparoscopy, and potent antimicrobial agents may actually be more pecuniarily sound than it appears at first glance.

10

Viral Infections in Obstetrics and Gynecology

Numerous viral diseases have relevance for the obstetrician and gynecologist, but currently three disease entities are of particular interest and will probably continue to have a major impact on the specialty over the next decade. The three viral agents discussed in this chapter, herpes simplex virus (HSV), human immune deficiency virus (HIV), and human papillomavirus (HPV), represent significant public health problems. Although virologists and infectious disease specialists are actively researching these conditions, the obstetrician-gynecologist frequently is the primary-care physician whose practice is directly affected by women infected with these agents. Brief attention is given also to parvovirus because of current interest in this etiologic agent. This chapter is not intended to exhaustively review all of the molecular biology of these viruses, but to selectively present material that may be of greatest use to the obstetrician-gynecologist.

HERPES SIMPLEX VIRUS

Overview

HSV belongs to the Herpesviridae family of viruses, which contains several specific viruses that not only have an impact on human heath but also specifically interest the obstetrician-gynecologist. These include HSV types 1 and 2 (HSV-1 and HSV-2), which cause genital and nongenital ulcerative infections and may cause severe and possibly fatal systemic infection in infants born to mothers who carry the virus.

A second virus that also has the potential for causing perinatal infection of the newborn is cytomegalovirus, which is responsible for mild symptoms in the infected mother. In affected infants symptoms may range from none to stillbirth or fulminant disease with jaundice and hepatosplenomegaly. Infants also may be born apparently normal, with the virus causing long-term problems with mental development.

Varicella-zoster virus is the causative agent of chicken pox and is of occasional concern because of its ability to cause a very severe form of pneumonia during pregnancy. The last virus belonging to the Herpesviridae family that is significant in human infections is the Epstein-Barr virus, which causes infectious mononucleosis, Burkitt lymphoma, and nasopharyngeal carcinoma.

Although the concern voiced by physicians, epidemiologists, and the general population about genital herpes infections has been eclipsed by the current interest in acquired immune deficiency syndrome (AIDS), the fact remains that about 500,000 new cases of genital herpes are estimated to occur each year in the United States, and the disease still is a public heath problem with ramifications for neonates.

Viral Agent

All herpesviruses are enveloped viruses with icosahedral nucleocapsids, and the viral genome consists of double-stranded DNA. The replication cycle involves the entry and uncoating of the virus in the host cell nucleoplasm, with production of new virus particles over a period of about 36 hours. The virus particles become enveloped during the process of egress through the nuclear membrane, are transported through the cytoplasm, and are released by passage through the golgi body or by exocytosis through the cytoplasmic membrane.

Although there are herpesviruses that affect almost all classes of warm-blooded animals and even some that infect fish and frogs, the human host is the only natural reservoir of HSV. There are two recognized types of virus, HSV-1 and HSV-2. Early epidemiologic observations indicated that HSV-1 infections usually involve the oral cavity and are transmitted by oral secretions, whereas HSV-2 was usually transmitted by sexual contact and caused genital lesions. However, the distinction has somewhat blurred in past years with the recognition that either virus type may infect the oral or genital mucosa. Other sites of infection, including the skin, are possible, and disseminated infection occasionally occurs, which indicates that there is no absolute restriction of the virus to oral or genital sites.

Clinical Infection

One of the gynecologic problems frequently seen in office gynecology practice is genital herpes or recurrent genital herpes. This infection will sometimes be asymptomatic, but when a patient presents with overt disease, the symptoms can be quite severe. The primary infection is manifested 2-7 days after exposure and is characterized by painful vesicular and ulcerated lesions that may involve the cervix, vagina, vulva, nearby skin, or urethra. The infection may involve the inguinal lymphatics and may occasionally become disseminated, causing a variety of complications including viral meningitis, usually in individuals who have compromised immune systems. Usually the virus reaches nerve endings, where it becomes latent until reactivated.

The exact sequence of events that triggers reactivation of a latent herpetic infection is unknown, but some factors that seem to be antecedents of recurrent disease have been noted. These include stress, sunlight, fever, or injury to the area of reactivation. Reactivation of a previous infection occurs when the virus begins to replicate in the chronically infected sensory ganglia. The virus migrates to a superficial site, where lesions form. Usually the symptoms are milder with recurrent disease. In primary infections the local symptoms usually persist for approximately 7 days, whereas recurrent lesions run their course in 3-4 days. Viral shedding may occur even in the absence of a visible ulcer, although active lesions appear to shed more virus.

Infection of the neonate is one of the most important aspects of herpesvirus infections for the obstetrician. Infection of the neonate with HSV can result in serious, often fatal, disseminated disease, because the infant seems not to have the necessary host defenses to control the spread of the virus. The infant most commonly is infected perinatally, and the majority of these infections involve HSV-2. Infection of the newborn may also occur as a result of contact with family members who are shedding the virus from oropharyngeal sources. Contact of the baby with infected oropharyngeal secretions should be as assiduously avoided as contact with infected genital secretions.

The infant infected in utero may be born with systemic disease or manifest systemic disease soon after birth. Transplacental infection, however, is not the usual source of perinatal infection. The virus reaches the fetus from the infected genital tract by ascension of the virus after membrane rupture. Alternatively, the infant may be infected during passage through the birth canal if virus is being shed. Perinatal infection may produce neonatal symptoms from birth to approximately 4 weeks. Disseminated skin lesions may be the most obvious sign of perinatal infection, occurring in about 70% of infected infants. Other signs of infection include irritability, jaundice, hepatosplenomegaly, seizures, chorioretinitis, pneumonitis, and bleeding disorders. Coma and death may result from respiratory distress and circulatory collapse. Some cases of neonatal disease are localized to skin, eye, or mouth and carry a better prognosis than disseminated infections. Mortality is about 70-80% for disseminated infection and approximately 30% for localized infection. Some reduction in mortality is possible with antiviral therapy.

Clinical Management

Discussion of the use of antiviral agents will be deferred until chapter 13, and nonchemotherapeutic aspects will be discussed here. When a mother is secreting virus, the major challenge for the obstetrician is to prevent the neonate from being exposed to the virus. Because the virus does not appear to cross the placenta or the intact fetal membranes, cesarean delivery has been found to be a rational way to prevent contact of the infant with the secretions of the lower genital tract in women actively secreting virus. Obviously, women with active lesions may be considered to be secreting virus. When commencing labor, these women will be candidates for abdominal delivery, as will be women who have ruptured membranes and active herpetic lesions. However, some disagreement exists as to the maximum length of time membranes may be ruptured and the infant still obtain a benefit from cesarean delivery. Usually it is considered that cesarean delivery should commence within 4-6 hours of membrane rupture; however, even with longer lengths of time after rupture, the theoretical benefit of abdominal delivery outweighs the risks.

The more complicated issue is that of women who do not have lesions at the time of labor or membrane rupture. The genital shedding of herpesvirus in asymptomatic women is estimated to be less than 0.5%; among women with a history of recurrent HSV infection, asymptomatic shedding may range from 8-14%. However, it has been estimated that the risk of virus shedding on the day of labor is 1.4%. In addition, the number of virus particles produced during virus excretion in the absence of lesions is smaller than the number produced by an individual with lesions. Thus, the asymptomatic individual may transmit disease less effectively than the symptomatic individual, although this remains to be proved and should not be a source of complacency for physicians caring for pregnant women who have a history of genital herpes. From the foregoing, it may be concluded that unless lesions develop near the time of expected delivery, it is reasonable to plan to have women with a history of recurrent infection deliver vaginally. This results in an estimated risk of neonatal exposure of 0.1%. When such individuals are in labor, a rapid nonculture (enzyme-linked immunosorbent assay [ELISA]) diagnostic test may be performed to determine whether cesarean delivery is warranted. Otherwise, culture of the mother and baby or a nonculture diagnostic test should be performed at the time of delivery to permit timely intervention with antiviral therapy if abdominal delivery is not undertaken.

A different situation exists when a woman has a genital lesion near the time of delivery, but the lesion heals before labor commences or membranes rupture. There is a reasonably high likelihood that such an individual will be a virus secretor at the time of delivery and should be cultured at 3- to 5-day intervals until delivery or be evaluated by a nonculture diagnostic technique to permit a last minute decision regarding cesarean delivery.

Diagnostic Considerations

In most instances, the presence of painful genital ulcers will be the source of suspicion. Although the lesions are fairly typical both in terms of appearance and painfulness, diagnosis should not be based on clinical observation alone. Culture is the best confirmation of an HSV infection, and because the virus replicates rapidly, culture results may be available within 24-48 hours. The physician should become aware of the laboratory capability and procedure in regard to herpes culture or nonculture testing before using such services.

Recently, antigen detection assays have been marketed. They may improve the ability to detect patients who are secreting virus at the time of labor or membrane rupture, although information supporting the validity of the use of these tests in the management of pregnant patients is not available at present. At least one ELISA is available that detects both HSV-1 and HSV-2 and has been approved by the Food and Drug Administration (FDA) for use without culture backup. Unfortunately, as with all antigen detection systems, the infectivity of the agent cannot be established with the immunodiagnostic tests. Undoubtedly, future work will attempt to evaluate the relationship of the presence or quantity of viral antigen to the likelihood of disease transmission. At present, any HSV antigen detected in the genital tract of a woman in labor would be considered a significant finding.

Prevention

As suggested by the foregoing discussion, the main mode of prevention is avoiding contact with the organism. In the case of HSV-1, this is almost impossible, as indicated by the fact that most adults have serologic evidence of exposure. Although exposure in later life is almost inevitable, neonates and immunocompromised individuals should be protected from exposure to persons with oral lesions. Fortunately, genital infection is not as prevalent as oral infection. But infection at either site is a risk for the infant, and every effort must be made to prevent infant contact with the organism by the means previously described.

The availability of vaccines for HSV-1 and HSV-2 will remain problematic for some time to come. Serum antibody is present after primary infection, but disease still reoccurs despite the antibody, suggesting that serum antibody is not protective. Moreover, there is great concern about using any killed vaccine because of the oncogenic potential of the herpesviruses. We understand from modern molecular biology that DNA even from a killed organism may recombine with host cell DNA and transform the host cell. Thus, there is at least a theoretic oncogenic risk even with inactivated virus.

HSV and Cervical Cancer

Because other members of the herpesvirus family definitely are associated with neoplasia, investigators have sought an association between HSV-2 infection and cervical cancer. Various case-control studies have indicated an association between HSV-2 antibody and cervical cancer, but very carefully matched case-control studies did not confirm the association. It is known that herpesvirus may transform tissue culture cells in vitro, and some tumor tissues have been found to contain HSV-2 DNA. However, in clinical observations, HSV may simply serve as a marker for sexual activity, and another agent such as HPV may more often be the factor most directly responsible for the tendency to develop cervical cancer.

HUMAN IMMUNE DEFICIENCY VIRUS

Overview

HIV infection is one of the most complex and challenging infectious disease problems to face science and medicine in this century. The care of patients infected with the virus often falls upon the internist, although because of the long latent period before symptoms are evident, many health care professionals will provide care to AIDS patients. Of special concern to the obstetrician is the efficiency with which the virus crosses the placenta, resulting in infection of the offspring.

Viral Agent

HIV is an enveloped, RNA-containing retrovirus that has been isolated from human populations in two forms: HIV-1 and HIV-2. Formerly used names have included HTLV-III (human T-cell leukemia/lymphotropic virus) and LAV (lymphadenopathy-associated virus). HIV-1 is the most common type in the U.S. population, with only a few cases of HIV-2 infection being reported compared with thousands of isolations of HIV-1. HIV belongs to a family of RNA tumor viruses that are also described as retroviruses because they encode the information required to synthesize DNA from an RNA template by means of the reverse transcriptase enzyme.

Much of what is known about AIDS with respect to pathogenesis, immune response, and potential methods of prevention would not be possible without a detailed knowledge of the structure and composition of the virus particle. Figure 10-1 shows in schematic form the arrangement of the components of this virus. As noted before, the genetic material of HIV is RNA, which resides within its core. Also colocated with the RNA is the reverse transcriptase enzyme. This enzyme is unique to retroviruses; without it they could not copy their genomes into a DNA transcript that then can be integrated into the host DNA. The virus also contains the core proteins p24 and p25.

The viral envelope is the first part of the virion that interacts with host cells and hence confers host and tissue specificity. The envelope consists of a typical membranous structure that is derived during the process of viral budding from the host membrane. However, the viral envelope contains glycoprotein structures that are responsible for many of the unique properties of this virus. One of these is a transmembrane glycoprotein, gp41; the other, gp120, resides outside the viral envelope in association with gp41.

The gp120 component is of great biologic interest because it binds to the CD4 receptor that is found on T-helper lymphocytes. In the simplest of terms, it is the ability of HIV to preferentially bind to and subsequently infect helper lymphocytes that causes disruption of normal immune function. This chapter will not attempt to detail all of the current thinking regarding what is probably a very complex interaction between the virus and CD4-bearing cells, but briefly, the viral core seems to enter the host cell by the formation of a syncytium between the viral envelope and the host membrane. Both gp120 and gp41 appear to play a part in this process, possibly by gp120 bringing the virus and cell together and the more hydrophobic gp41 causing fusion of viral and cell membranes.

Although most conventional viruses use the host cell's biosynthetic capability to produce new virus particles, the AIDS virus forms a DNA copy of its genome that is integrated into the host genome in a lysogenic state. Here the viral genes may either not be expressed at all or be expressed by production of small numbers of virus particles with minimal cell damage. Alternatively, the host cell may be altered in such a way that it forms syncytial complexes with other cells, resulting in the death of the cells, or the HIV may kill the cells by other less obvious means.

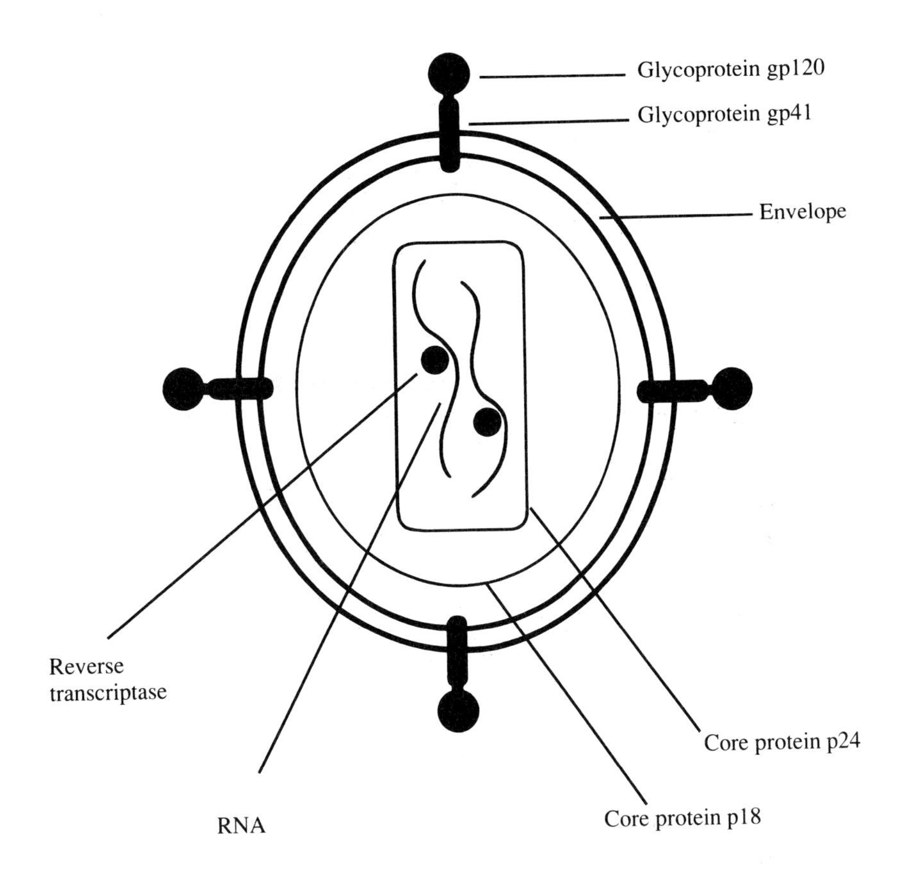

Fig. 10-1. *Physical structure of the AIDS virus.*

The specific manner in which HIV affects cells remains a subject of intense investigation. It appears that there may be multiple mechanisms whereby cells are damaged or altered by HIV. Moreover, T-helper cells are not the only cells infected, which further complicates the understanding of all of the details of the virus-cell interaction. A significant proportion of peripheral blood macrophages and a smaller proportion of B lymphocytes have CD4 on their surfaces. Glial cells of the brain and chromaffin cells in the gut are also susceptible to HIV infection, but such interaction may not depend entirely on the presence of CD4, because these cells produce little or no CD4 marker.

Also important is the fact that the components of HIV are immunogenic, and individuals infected with the virus do produce antibodies to these antigens. This is an important aspect in the diagnosis and clinical course of the disease and will be further discussed in the section "Diagnostic Considerations."

Source of Infection

As suggested by the preceding discussion, the AIDS virus may be found in various tissues; viral cultures have demonstrated that it is present in many body secretions as well. These secretions include blood, saliva, semen, tears, amniotic fluid, breast milk, vaginal secretions, and cerebrospinal fluid. However, it is absolutely essential to recognize that not all of these are epidemiologic sources of infection. The infection appears to be transmitted mainly through sexual intercourse or through blood contamination and less commonly by breast milk.

The blood-borne virus has at least two major implications for the obstetrician-gynecologist. First, the organisms in maternal blood may be transplacentally transmitted to the fetus, resulting in the birth of a baby with AIDS. Second, because obstetrics and gynecology is a surgical specialty, the woman infected with HIV presents a risk to the surgeon and operating-room personnel. This risk must not be overemphasized, but should primarily serve as a reminder that the physician and assistants must avoid contaminating themselves with blood. This precaution has always been an important part of medical practice because of the risk of syphilis and hepatitis B infection, but now, more than ever, the additional risk of AIDS has introduced more stringent protective measures into medical practice. Each hospital's epidemiology/infection control hierarchy should have established guidelines for providing care to known or suspected AIDS patients, and operating rooms should have the necessary standard procedures and necessary protective garb.

The risk of intrauterine AIDS clearly is associated with population groups or geographic centers where the prevalence of AIDS in the adult female population is high. Because the disease in the U.S. population involves a significant homosexual component, the male/female ratio has remained relatively high since the disease was discovered. In Africa, where heterosexual transmission is common, the male/female ratio is near 1. Thus, the women at greatest risk in the United States now appear to be those who engage in intravenous drug use or who are partners of men who are at high risk for AIDS. Seroprevalence studies among infants born in inner-city hospitals in areas of high prevalence have shown that 1-4% of infants have serologic evidence of infection. This is, of course, an upper limit because these populations are probably at the highest risk for AIDS. The pregnant patient may also experience some modulation of her immune responses as a result of physiologic adjustments to pregnancy (see chapter 2), and it is unknown whether this may influence the course of HIV infection during pregnancy.

Clinical Disease

HIV infection may occur with little immediate clinical consequence except for mild flu-like symptoms that may not even be noticed. The complex of symptoms indicating AIDS may not appear until years after the virus is first acquired. The clinical disease in adults is a heterogeneous entity. Not all patients have a single specific set of symptoms. Furthermore, because the disease progresses over a long course of time, not all symptoms are likely to be present during the early stages of the disease. Immunodeficiency is manifested by opportunistic infections. Although one of the most notorious is *Pneumocystis carinii* pneumonia, which served as part of the original case definition, the list of other opportunistic infections that may occur in a patient with AIDS is exceedingly long.

A reasonable description of the clinical and pathologic consequences of AIDS is provided by the Walter Reed Classification of the disease (Table 10-1). These criteria allow the staging of AIDS from exposure through subclinical phases into the stage when clinical symptoms are apparent, and finally into the last stage to which the term AIDS is appropriately applied. The stages are identified as WR0 (exposure) through WR6 (AIDS). If Kaposi sarcoma is present, it is appended to the stage (for example, WR6K). A patient enters WR1 when the virus is present or antibody tests are positive for HIV. WR2 is heralded by chronic lymphadenopathy, and WR3 is characterized by a decline in T-helper cells to below 400/μL. WR4 is signalled by loss of response to at least one of four skin-test antigens, with the anergy persisting for at least three months. The next stage is characterized either by oral candidiasis or by loss of reactivity to all skin-test antigens. Finally, WR6 is indicated by the presence of opportunistic infections.

The symptoms of HIV infection and AIDS in infants may differ somewhat from that observed in adults. An embryopathic effect has been claimed in pregnancy complicated by AIDS. The disease may progress quickly in infants, as suggested by findings from New York City that indicated that 50% of AIDS babies did not survive beyond 15 months and only 15% survived to the age of 5 years. AIDS babies have reduced responsiveness to antigens, including those in vaccines. They may, therefore, be especially prone to a wide variety of bacterial infections, even more so than adults with AIDS. Although adults commonly acquire *Pneumocystis carinii* pneumonia, AIDS babies are more likely to have lymphoid interstitial pneumonitis. Kaposi sarcoma is less common in pediatric AIDS patients than in adults with AIDS, but lymphoma and other lymphoreticular cancers are more likely. Neurologic problems also occur in children as a result of AIDS and typically reflect interrupted or diminished brain growth and intellectual development.

Diagnostic Considerations

HIV may be cultivated in cell culture, although relatively few laboratories provide this as a diagnostic test. Despite the fact that HIV infection results in immunosuppression, it is fortunate that an antibody response occurs after contact with the virus. As indicated

Table 10-1. *Walter Reed Staging Classification for HIV Infection*

Stage	Criteria
WR1	HIV antibody or virus isolation
WR2	Plus chronic lymphadenopathy
WR3*	Plus T-helper cells ‹ 400/mL
WR4	Plus partial cutaneous anergy
WR5	Plus complete cutaneous anergy to tetanus, trichophyton, mumps, and candida; and/or thrush
WR6	Plus opportunistic infections

*Lymphadenopathy may subside after WR2.

above, there are several presymptomatic stages of HIV infection before AIDS is manifested clinically; likewise, several years may intervene between exposure and clinical disease. Therefore, early detection through the use of immunodiagnostic tests may permit the use of antiviral therapy that in turn may prolong the patient's life.

Within 2 weeks of exposure, viral antigens including p24 and gp41 are present in the serum, and antibodies to these antigens begin to develop about 6 weeks after exposure. Numerous tests have been developed to examine both antibody to HIV and the antigens of the organism, although each test has limitations and not all are FDA approved.

Initial detection of antibody against HIV usually is done by an enzyme-linked immunoassay. Several companies produce materials for antibody detection. Obviously, such assays are not without problems of false-negative and false-positive results, necessitating the use of supplemental tests. Western blot analysis remains the mainstay of diagnosis. The advantage of Western blot is that it detects antibody against several components of the virus. However, the test is technically demanding, and results are not always unequivocal. Researchers are still debating the appropriate interpretation of indeterminate Western blots.

Testing of the patient's serologic status should be accompanied by a thorough history of known or presumed exposure to an infected individual, the presence of risk factors such as intravenous drug use or having a partner who is at high risk for acquiring AIDS, physical examination, and proper counseling. In addition, evaluation of the ratio of T-helper to T-suppressor cells (T4/T8) may also be of value.

An infant suspected of having AIDS presents a further problem in diagnosis, because immunoglobulin G (IgG) antibodies cross the placenta and if cord blood is positive for HIV antibody, it may represent maternal antibody. Western blot specific for IgM antibody may be an appropriate method, although the availability of this technique is limited.

HUMAN PAPILLOMAVIRUS

Overview

Two facts about this virus have thrust it into the forefront of concern among gynecologists. First, techniques of modern molecular biology have revealed that its prevalence is greater than previously realized. Second, it has been strongly associated with cervical neoplasia and may consequently be one of the most significant causes of cervical cancer.

Viral Agent

HPV is a nonenveloped DNA virus that has not been cultivated in cell culture. The method for characterizing papillomaviruses is evaluation of the patterns of DNA fragmentation after restriction endonuclease treatment and DNA hybridization. On the basis of DNA homology, more than 50 types of papillomavirus are known, and many of these may infect the female genital tract. In general, HPV infections involve only the upper epithelial layers. Types 6, 11, 16, 18, 31, 33, and 35 are most often associated with the genital infections. Other types are associated with common skin warts.

Source

It is generally conceded that genital HPV infections arise after contact with an infected sexual partner. Observation of exposed individuals indicates that the development of genital condyloma occurs 6 weeks to 3 months after exposure. In addition, infants born to mothers who have genital infection may subsequently develop laryngeal papilloma. It has recently been recognized that it is possible to be infected with multiple types of virus simultaneously.

Clinical Infection

To some extent the clinical manifestation of HPV infection is determined by the viral type. Exophytic warts are commonly associated with type 6 or 11, although lesions on the cervix are usually flat or depressed, regardless of type. Condyloma may be found on the vaginal walls and the vulva, or they may be located perianally. They have been noted to increase strikingly in size in response to pregnancy.

Diagnostic Considerations

Colposcopy is an important aspect in the diagnosis of the infection, and blanching of the lesions after application of acetic acid is an important tool. The same technique may be applied to the penis of the male partner. Because of the sexually transmissible nature of this condition, coupled with the association between HPV and the development of dysplasia, physicians must take more responsibility for ensuring that the male is evaluated and treated as far as is practicable.

The presence of koilocytes in the lesion is pathognomonic. If the laboratory has the capability, typing the virus is useful to assess the potential aggressiveness of the lesion. In any case the patient must be informed that attention to follow-up appointments is necessary to mark the progress of the disease, if any. Given the possibility of serious sequelae resulting from HPV infection, objective methods of diagnosis and strain typing are needed. Currently DNA-hybridization assays including Southern blot, dot blot, or hybridization of gene probes with fixed or frozen tissue samples are utilized. However, these techniques are mainly used in research centers, and adequate standardization of gene probes is lacking. The advent of the polymerase chain reaction for probing infected tissues appears to be a method that potentially will provide exceedingly sensitive detection of the viral genome, although identifying the site where sampling should be done remains problematic because tissues may be infected before lesion formation. Because there are no antiviral drugs useful for HPV infection, ablation of the lesions is the mainstay of therapy. The various techniques available will not be reviewed here.

Predisposing Factors

The majority of demographic and behavioral risk factors are those typical of sexually transmitted diseases. Factors that increase the risk of infection also appear to increase the risk of developing cervical cancer. Multiple sexual partners, sexual experience at an early age, and having a male partner previously married to a woman with cervical cancer have been noted as risk factors for the development of HPV infection.

Oncogenic Potential

It is not entirely clear how HPV causes malignant transformation of cells. Viral DNA may be present in cells separate from the host chromosomes with 5-200 copies of the viral genome per cell, or varying amounts of the viral genome may be integrated into the host DNA. It is believed that every infected cell is transformed, although not all resultant tumors are malignant. The progression to malignancy occurs in the host at different rates depending on various factors, including the virus type and probably certain host factors, as well as cofactors. Smoking, diet, and hormonal conditions have been named as possible cofactors in carcinogenesis related to HPV.

A large body of literature is available describing the evidence for HPV involvement in cervical cancer, but no attempt will be made here to recount these studies. Briefly, one of the most striking findings is that virtually 100% of cervical biopsies obtained from precancerous lesions contain HPV DNA. In addition, higher-grade lesions are more likely than lower-grade lesions to contain DNA from HPV type 16. HPV type 16 DNA is most consistently found in tumor tissue samples, although types 18, 31, 33, and 35 have also been identified. Continuously replicating cell cultures derived from cervical carcinomas (HeLa, CaSki, and SiHa) contain conserved DNA sequences from HPV. Despite these striking associations, it must be noted that type 16 DNA may be found in cells from women without precancerous changes. Because of the difficulty in obtaining meticulous matches in case-control studies, it may be some time before the true nature of the association is known.

PARVOVIRUS

Numerous viral agents that have not been mentioned in this chapter have the potential to cause infection in women and may also complicate pregnancy. An exhaustive description of these illnesses will not be included here; however, it is appropriate to note that among the emerging problems facing the obstetrician, parvovirus infection is gaining recognition as a perinatal pathogen. Most individuals who acquire parvovirus infection are infected intranasally by means of droplets, although parvovirus has been reported to have appeared in blood products. Despite the fact that about one third of adults have parvovirus antibody, most individuals are unaware of the disease. Symptomatic infections are accompanied by a typical viral syndrome of fatigue, fever, malaise, arthralgias, arthritis, and a diminished hematocrit and leukocyte count. Occasionally the illness is accompanied by a rash described as erythema infectiosum or fifth disease. When parvovirus infection is acquired during pregnancy, fetal death due to anemia caused by hydrops fetalis may occur, but congenital anomalies are unknown.

At present there is no adequate diagnostic method for detecting this infectious disease, nor are there any proven therapeutic agents. Immunoglobulin preparations containing antiparvovirus antibody appears to be a potentially beneficial future therapy, whereas vaccination will ultimately be the most reliable future method of disease control. Fortunately, despite the prevalence of the disease, it is only infrequently incriminated in spontaneous abortions, although more rigorous surveillance could elevate its status as a perinatal pathogen.

Part IV

Medical Approaches to Infectious Disease in Obstetrics and Gynecology

11

Diagnosis of Gynecologic and Obstetric Infections

In this section, the microbiologic basis for the obstetrician-gynecologist's knowledge of infectious diseases is applied practically to the clinical infections that represent such a large but frequently ignored part of the specialty. Obviously, most infections do not present extraordinary diagnostic challenges. Although most clinical decisions regarding diagnosis and therapy of infections are not ponderous, it is valuable to review a logical approach to infection as it relates to the basic microbiology presented. This approach is essentially algorithmic and seeks to include the maximum amount of information in formulating a diagnosis.

SUSPICION OF INFECTION

Very basic considerations cause a physician to begin the process of doing an infection workup. A patient who presents with symptoms that might be attributable to infection is an obvious starting place in most instances. The patient who has undergone a surgical procedure will be monitored for objective signs of infection, and when these appear, diagnostic workup will begin. Finally, some patients who are asymptomatic will be evaluated for the possible presence of infection, including women who are sexual partners of men who have a sexually transmitted disease and women seeking routine care. The physician has the opportunity to discover symptoms unnoticed by the patient when she obtains gynecologic or prenatal care.

LOCALIZING THE SYMPTOMS

Physical examination and review of symptoms will usually lead the physician to identify the site of infection. In the case of the patient who has undergone surgery, attention naturally is directed first at the operative site. If no local evidence of infection is obtained, attention is turned to other sites of compromised host defense. As noted in earlier chapters, the host's local defenses are affected by many of the things that occur incidental to surgery, including endotracheal intubation, bladder catheterization, and placement of indwelling intravenous lines. These are the next sites evaluated for signs of sepsis. Systemic symptoms may be due to disseminated infection, although signs of bacteremia may indicate the seeding of the bloodstream from a nidus of infection.

The physician can use a variety of techniques to aid in the process of localizing the source of symptoms. In addition to physical diagnostic techniques, various imaging techniques, including ultrasound and magnetic resonance imaging, may be appropriate means of identifying an abscess and may also be useful in discovering noninfectious causes of symptoms. Gallium-67 or indium-111 labeling of white blood cell accumulations may also be done, although in many cases the physician will not need to resort to techniques with this degree of sophistication and cost. Endoscopic evaluation is also a tool that is becoming increasingly valuable, not only for pelvic inflammatory disease but also for other conditions as well. Recently, some surgeons have employed laparoscopy for draining abscesses as well as diagnosing them. However, the skill of the operator must surely be considered when contemplating entrance of an abscess cavity via laparoscopy.

Along with these clinical observations, laboratory tests other than culture may prove useful in helping to verify that the symptoms are indeed due to an infectious process.

The universal utility of the leukocyte count and differential is obvious. The erythrocyte sedimentation rate and elevated levels of C-reactive protein are associated with inflammatory processes and have been considered by some to be useful in diagnosis of pelvic inflammatory disease and chorioamnionitis. However, these tests are nonspecific and merely help to corroborate the presence of infection or an inflammatory process.

ESTABLISHING A WORKING DIAGNOSIS

Once the physician is satisfied that the symptoms observed are due to infection and has established the site of infection, the next priority becomes determining the etiologic agent. The range of microorganisms that may be responsible for infection at a given site are usually known as a matter of prior experience. As described in the preceding chapters, a reasonable prediction of which organism or organisms may be involved in the illness can usually be made, although the specific organisms present at an infected site can only be determined by culture and other specific identification methods.

It is important to know which microorganisms are likely causes of infection at a given site. However, in addition to having a textbook list of organisms, local experience may figure extensively in establishing the possible cause of infection. Certain types of microorganisms may be more common in one hospital than another, and certain resistance patterns among these organisms may be unique for a certain hospital. The value of the hospital epidemiologist, working in conjunction with the clinical microbiology laboratory, lies in the ability to provide information regarding prevalence of organisms and antibiotic resistance patterns. Such knowledge can influence the selection of antibiotics for empiric therapy and may be the deciding factor in choosing not to use a particular antibiotic, simply for the sake of avoiding any contribution to known locally developing resistance problems. Vaginal infections and sexually transmitted disease may show unique geographical distributions. For example, a clinician's thinking and approach to diagnosis will be influenced by the knowledge that his or her practice is in an area of high endemicity for human papillomavirus, or that more patients in that location have trichomonal vaginitis than yeast vaginitis.

At this time additional observations such as Gram stain of material obtained from the site of infection may further guide the clinician in determining the most probable cause of infection. Many potential sites have a normal flora that usually confounds the results of Gram stain or other special stains. Gram stains of normally sterile materials are most helpful, especially if the organisms of interest have a unique morphologic appearance. Thus, centrifuged urine, amniotic fluid, gastric aspirates from neonates, cerebrospinal fluid, and aspirates of abscesses may provide information about the presence of certain organisms. *Clostridia* and *Neisseria* are examples of organisms with a distinct morphology that may be apparent on Gram stain. In cases of vaginitides, direct observation is likewise essential for the identification of clue cells, yeast buds, and hyphae or trichomonads. In addition, Gram stain of vaginal material may reveal the curved gram-negative rods that have been seen in some cases of bacterial vaginosis.

The use of direct microscopic observation may be of further value in distinguishing bacterial from viral infections. Careful cytologic evaluation of cervical smears may reveal cytopathic changes consistent with certain types of viral infections.

It may seem inappropriate to invest time in attempting to obtain a presumptive diagnosis by clinical observation and direct microscopic observation. However, in some instances the plan of treatment may be influenced by these observations and in other cases, especially the vaginitides, microscopic observation is adequate to establish the etiology of infection and initiate therapy.

IDENTIFYING THE ETIOLOGIC AGENT

The gold standard for the diagnosis of most infectious diseases is culture. This statement applies to both bacterial pathogens and viral or chlamydial agents. Several practical limitations must be noted, however. Not all infectious agents are amenable to culture. One such noncultivatable bacterial pathogen is *Treponema pallidum*, and a viral agent that cannot be grown in cell culture is human papillomavirus. In addition, in the case of mixed bacterial infections, typically upper-tract infection in pelvic inflammatory disease and postoperative sepsis, it is not clear what the precise role of each organism is or which organisms are appropriate targets of antimicrobial therapy. Finally, culture takes time and may not be available soon enough to influence patient management. As a consequence, numerous specific non-culture-based techniques are in use or are being developed. This is obviously not a static field of endeavor, and any attempt to list all of the non-culture-based diagnostic products will soon be out of date.

Some of the oldest methods for diagnosis have relied on detection of serum antibody to the infectious agent. Syphilis is diagnosed by such a method, although the screening test is not specific. The confirmatory (FTA-Abs) test is specific. Viral infections are also frequently identified by serodiagnostic methods because viral culture requires specialized laboratory facilities that may not be available in all hospitals. AIDS testing is one example of diagnosis that is largely based on the patient's antibody response to the virus. Antibody detection methods may also be an important way to determine susceptibility to infectious agents such as the rubella virus.

Direct tests that detect viral or other antigens are also becoming important in clinical practice. Antigen detection of chlamydia in endocervical samples is not only possible, but has been adapted to a system that may be used in an office setting without the need for a complete clinical laboratory. As noted elsewhere, the interpretation of these tests by nontechnicians and the adequacy of quality control remain issues that will not soon be settled. Antigen detection for group B streptococcus is also available, and its role in allowing timely decision-making in the delivery room is still evolving.

An innovation that will have an immediate impact on diagnosis of perinatal infection is the availability of genetically engineered G-protein. This protein binds specifically to immunoglobulin G (IgG) and may be attached to a solid support, hence becoming the basis for an affinity chromatography column. Simply constructed, disposable columns may be used in conjunction with tests for antibody against specific organisms. In diagnosis of fetal infection, cord blood IgM specific for the suspected pathogen indicates acute fetal infection because IgM cannot cross the placenta and therefore must have originated with the fetus. Cord blood passed through a G-protein column will have the maternal IgG removed, leaving primarily the fetal IgM if it is present. This may then be used in specific antibody detection assays.

The physician should confirm the availability of individual diagnostic tests by consultation with the laboratory. Some tests that may not be locally available might be accessible through state, county, or municipal reference laboratories or through a university-based medical center. It is advisable to determine the specimen requirements for the desired tests, methods for handling and shipping, and expected turnaround time.

The most promising of the rapid laboratory tests available for specific detection of microbial pathogens is based on gene probe technology. Gene probes detect specific sequences of microbial genetic material and are of interest not only because of their exquisite specificity but also because they can detect genetic sequences of viruses that have become lysogenized into the host genome. This technology is now being applied to such viral agents as human papillomavirus. Not only is it possible to identify cells that carry the virus in a latent state, but it is also possible to determine the viral type, an important aspect of establishing a prognosis for the oncogenic potential of a lesion.

Although genetically engineered test reagents may in the future represent the majority of diagnostic methods, for the immediate future, microbial culture methods will continue to play a major role in diagnosis of infectious diseases. Because of this fact, the physician must remain cognizant of the importance of specimen collection for specific types of microorganisms. This reliance on microbial culture methods further emphasizes the importance of making a working diagnosis so the process of selecting an appropriate culture method may be optimized. In the case of certain fastidious organisms, such as *Neisseria gonorrhoeae*, the organism is placed directly on the bacteriologic growth medium by the physician and placed immediately into an appropriate CO_2-enriched atmosphere. Other organisms, such as *Haemophilus ducreyi*, may also require direct plating on special media.

The physician should work closely with the laboratory if nonroutine cultures are to be done. The laboratory may provide special instructions regarding handling of the specimens. Likewise, the way the laboratory handles specimens may be influenced by the organisms that the physician suspects or wishes to rule out. *Listeria* isolation is aided by the cold enrichment technique, but this is not routinely done to cultures that are not specifically identified. If *Mycobacterium tuberculosis* is suspected, the laboratory will handle the culture in a laminar-flow protective hood to prevent aerosol contamination of the technician. *Campylobacter* is rarely present in obstetric and gynecologic specimens, but because it requires isolation on special media and incubation in an enriched atmosphere, the laboratory should be notified if this organism is suspected. The importance of maintaining anaerobiosis for specimens that may contain obligate anaerobes always bears repeating. Urine cultures must not be allowed to remain at room temperature, and specimens collected for urine culture should not have antibacterial preservatives added.

FOLLOW-UP

The diagnostic exercise often continues beyond the point at which a specific microorganism is incriminated in a disease process. In most cases therapy is given and the patient improves uneventfully. Occasionally "test-of-cure" cultures may be desired, as in the case of *Neisseria gonorrhoeae* or *Chlamydia trachomatis*. In the case of serologically diagnosed diseases such as syphilis, declining antibody titers may be part of the posttherapy surveillance routine.

In some cases the illness does not improve or appear to improve, or some new problem emerges. For example, patients receiving antibiotic therapy may develop diarrhea, which may indicate that antibiotic-associated enterocolitis has supervened. The process of diagnosis then begins again, with the confirmation that the stool contains enterotoxin, and new therapy is undertaken. Herpetic lesions and other ulcerative lesions may become secondarily infected by lower-genital-tract microorganisms and represent new disease entities that require a different type of treatment. In its early phases, intraabdominal infection may be primarily a soft-tissue infection, but in some individuals it will give rise to an abscess that requires surgical attention. Patients receiving antibiotic therapy have developed toxic shock syndrome when the toxic shock strain of *Staphylococcus aureus* was not sensitive to the antibiotic originally given. In each of the examples presented, the secondary complication must be recognized not as an inadequately resolved primary infection, but as a new diagnostic problem that may require a new therapy. This underscores the necessity of careful reevaluation of all patients who do not have appropriate resolution of symptoms or who seem to have recrudescent disease.

In the subsequent chapters, the emphasis will be placed on rational selection of therapy based on sound microbiologic concepts. The same orderly and rational attention to the microbiologic concepts presented previously in this monograph is required to provide the proper diagnosis before the best therapy may be selected.

12

Chemotherapeutic Choices

An Algorithm for Selection of Antimicrobial Agents
Infectious Diseases and Therapy

- Vaginal Infections
- Toxic Shock Syndrome and Staphylococcal Infections
- Mixed Infections After Gynecologic Surgery
- Postpartum Endometritis and Infection After Cesarean Delivery
- Infection Complicating Malignancy
- Bacteremias
- Bacteriuria
- Sexually Transmitted Diseases
- Viral Infections

AN ALGORITHM FOR SELECTION OF ANTIMICROBIAL AGENTS

In this chapter, practical aspects of antibiotic use are discussed. Previously the microbiologic implications of antimicrobial agents were discussed, but here the relationship to specific infectious disease entities will be considered. It must be emphasized again that this volume is not intended to be a collection of protocols for managing infectious diseases. To underscore that fact, this chapter on the clinical use of antimicrobial agents has been kept near the back of the book. Also in keeping with the emphasis on basic science principles, this chapter on selection of antimicrobial agents will not provide specific recommendations or dosages.

The selection of antimicrobial therapy will ideally involve the use of a drug that has been selected by a rational process from hundreds of options. An algorithm that presents the decisions that are made in rationally selecting an antimicrobial agent is shown in Fig. 12-1. In clinical practice these choices are far more direct than might seem to be implied by the decision tree. Nevertheless, all of these elements play a part in choosing therapy.

The inclusion of cost in the decision tree is warranted in view of the current climate of cost consciousness and medical cost containment. Fortunately, it is possible to make decisions that will save money without jeopardizing the level of care provided or the efficacy of the therapy selected.

INFECTIOUS DISEASES AND THERAPY

The preceding chapters presented a summary of the microbial diseases that are of primary importance to the obstetrician-gynecologist. Obviously, as the primary-care physician of women, the gynecologist encounters many other infectious disease entities that are unrelated to the reproductive tract. However, these other infectious disease entities will be excluded from this chapter. Only the conditions noted in chapters 5-10 will be covered here, with some comments on therapeutic choices that are consistent with the principles of the host-antibiotic-microorganism interaction. The reader is reminded that the discussion of antimicrobial agents below should not be construed as indicating that antibiotics alone represent the sum of appropriate therapy for these infections. It is

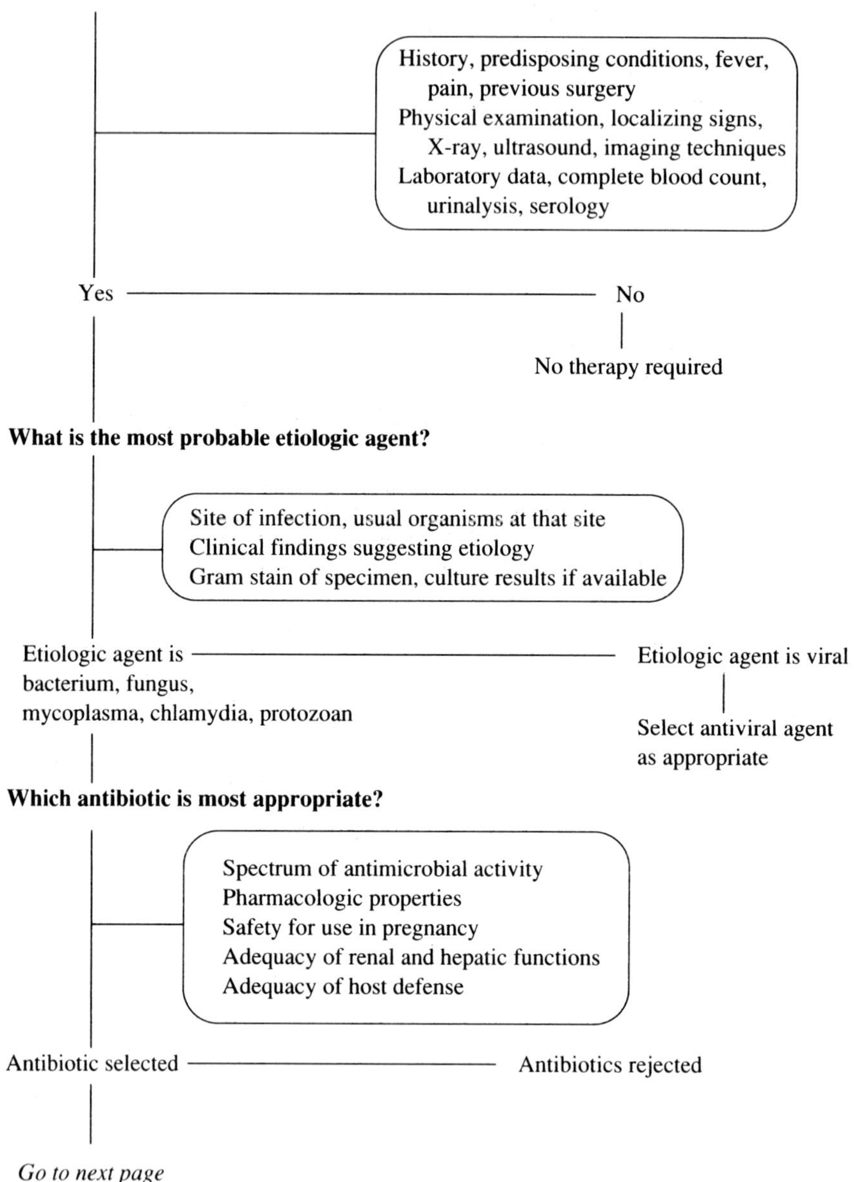

Fig. 12-1. *Algorithm for selecting an antimicrobial agent.*

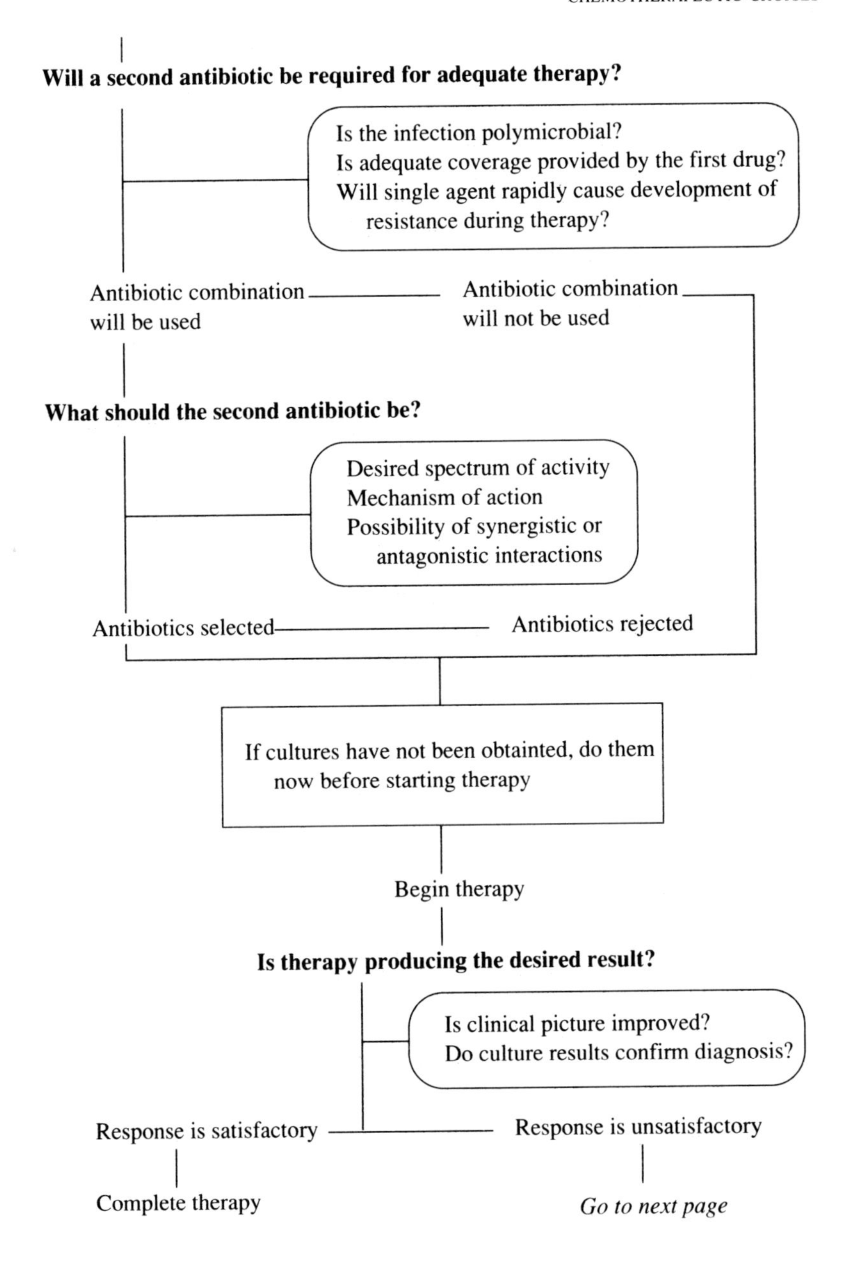
Will a second antibiotic be required for adequate therapy?
Is the infection polymicrobial?
Is adequate coverage provided by the first drug?
Will single agent rapidly cause development of resistance during therapy?
Antibiotic combination will be used
Antibiotic combination will not be used
What should the second antibiotic be?
Desired spectrum of activity
Mechanism of action
Possibility of synergistic or antagonistic interactions
Antibiotics selected
Antibiotics rejected
If cultures have not been obtainted, do them now before starting therapy
Begin therapy
Is therapy producing the desired result?
Is clinical picture improved?
Do culture results confirm diagnosis?
Response is satisfactory
Response is unsatisfactory
Complete therapy
Go to next page

(Continued)

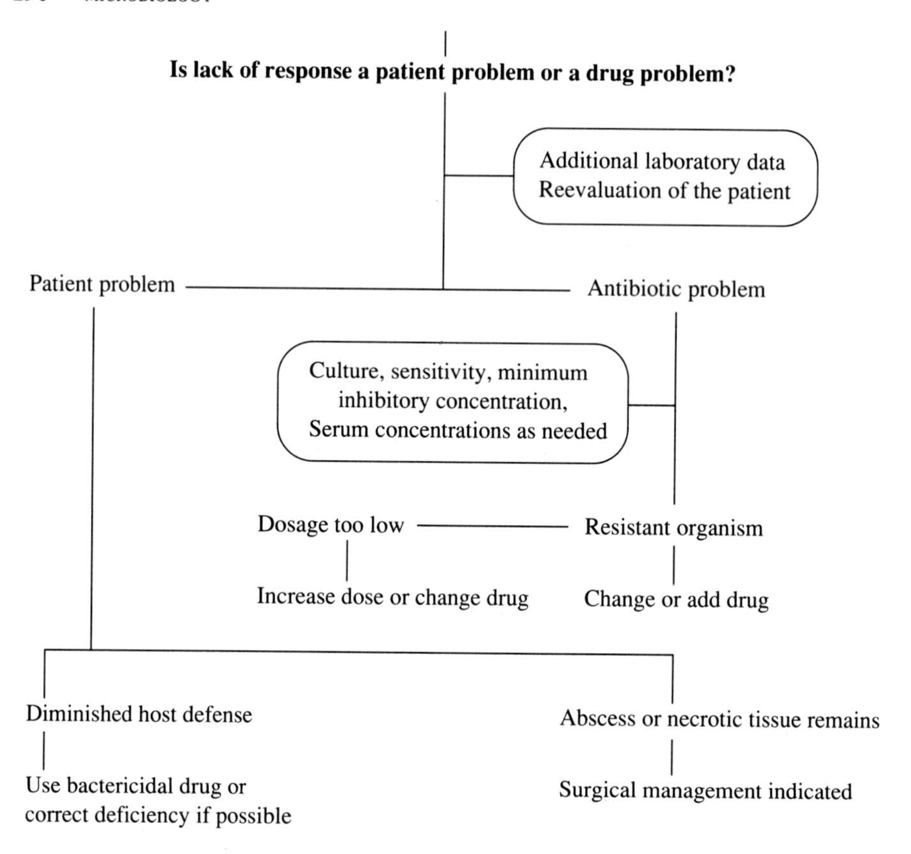

Fig. 12-1. *Algorithm for selecting an antimicrobial agent (continued).*

assumed that all other necessary measures, such as drainage of fluctuant masses, supportive therapy for shock patients, nutritional support, and other measures appropriate to the specific condition, will be provided.

Vaginal Infections

The common vaginal infections represent one of the most frequent opportunities for the gynecologist to make an astute, well-reasoned diagnosis and to provide appropriate therapy. The role of accurate diagnosis in providing satisfactory therapy cannot be overemphasized.

CANDIDA ALBICANS

Candida albicans usually responds to topical therapy with one of the triazole compounds such as miconazole, terconazole, butoconazole, or clotrimazole, with therapy continued for 3 days. Various dosage forms, including intravaginal tablets, suppositories, and creams, are available. Physician experience and patient acceptance are the major aspects in selecting a dosage form. Some patients will need a dosage form that will allow application to affected vulvar areas, as well as intravaginal sites. Patients should be made to understand the importance of continuing the medication for the recommended length of time to avoid relapse.

BACTERIAL VAGINOSIS

Bacterial vaginosis is now known to involve anaerobic bacteria; as a consequence, the therapy of choice is metronidazole. The drug is given in two oral doses of 500 mg each for 7 days. The patient should be warned about the possibility of a disulfiram-type reaction if alcohol is ingested. An alternative therapy is oral clindamycin given in 300-mg doses twice daily for 7 days.

TRICHOMONIASIS

In the majority of cases, trichomoniasis is also treated with metronidazole, but because of the exquisite sensitivity of the organism, a single 2-g dose is therapeutic. Treatment failures can be a vexing problem. The primary approach has been to increase the dose of metronidazole, and in some cases to administer the drug intravenously in high doses with due regard for the potential toxicity of such regimens.

Toxic Shock Syndrome and Staphylococcal Infections

Staphylococcus aureus is one of the most frequently encountered organisms in skin and other infections. It has a propensity to produce abscesses that require drainage as part of therapy. The most likely setting in which an obstetrician-gynecologist would see the organism is in postoperative wound infections, stitch abscesses, or in cases of surgical toxic shock syndrome.

Most menstrually related cases of toxic shock syndrome are likely to come under the care of emergency physicians, family practitioners, or internists. In selecting antibiotic therapy, it should be remembered that one of the characteristics of toxic shock strains of *Staphylococcus aureus* is production of penicillinase. Thus, penicillinase-resistant drugs such as nafcillin or oxacillin are considered appropriate choices. Alternative therapy for patients hypersensitive to penicillin includes vancomycin or cefazolin. Therapy should be continued for 14 days, but parenteral drugs may be replaced with oral medications when the patient is able to tolerate them.

Hospital strains of staphylococci are more likely than community strains to be penicillin resistant. Therefore, when infection of a surgical wound occurs, it is appropriate to consider the organism penicillin resistant until laboratory results are available. The physician should be aware of the prevalence of methicillin resistance among staphylococcal isolates from his or her institution because this may influence the choice of vancomycin. *Staphylococcus aureus* or *Staphylococcus epidermidis* may cause vasculitis at the site of indwelling venous catheters, and penicillinase-resistant penicillins or vancomycin may be used to treat these infections, along with removal of the catheter.

Breast infections that are monoetiologic are usually due to *Staphylococcus aureus* and may be treated with penicillinase-resistant penicillins or vancomycin. If their etiology is polymicrobial, they may respond to cefoxitin, which will provide coverage for anaerobic species that may be present. The physician should also be cognizant of the cost of vancomycin.

Mixed Infections After Gynecologic Surgery

The polymicrobial nature of infections arising as a complication of gynecologic surgery has been noted. From careful culture studies of infected patients and trials of various therapeutic regimens, the goal of therapy has become antibiotic coverage of gram-negative facultative organisms such as *Escherichia coli* and simultaneous coverage of the anaerobic organisms, particularly the *Bacteroides*. It is well known that the vaginal flora frequently contains enterococci, and providing antibiotic coverage for these organisms is considered important by some surgeons. Unfortunately, therapy of these types of infection is usually undertaken without the results of cultures, and so drug therapy is selected empirically. Because of the wide range of organisms that may participate in these infections, it is common for therapy to consist of two and sometimes three antibiotics. With the advent of the second-generation cephalosporins, monotherapy is also possible. Several possible therapeutic options are presented below.

The combination of clindamycin and gentamicin is often referred to as the gold standard because it has a long history of use and provides a high level of intrinsic activity against the organisms commonly found in mixed infections. Clindamycin provides anaerobic coverage, whereas gentamicin effectively inhibits gram-negative facultative rods. The local rate of resistance to gentamicin is an important consideration, and a different aminoglycoside such as amikacin may be employed. The combination of clindamycin and gentamicin does not provide coverage for the enterococci.

Monotherapy with a second-generation cephalosporin such as cefoxitin or the more recently introduced cefotetan is also an option. Some specialists have elected to use monotherapy when the infectious process is localized and multidrug therapy when the infectious process extends beyond the operative site. Drugs used for monotherapy of postoperative infection have activity against the gram-negative facultative bacteria and anaerobic organisms, although activity against anaerobes is probably slightly less than that of metronidazole or clindamycin. At a blood concentration of 16 mg/ml, cefoxitin inhibits about 70% of *Bacteroides fragilis* strains, whereas clindamycin inhibits 80% of the strains at a concentration of 0.4 mg/ml, and metronidazole inhibits 90% of the strains at a concentration of 4 mg/ml. The intrinsic antibacterial activity is not the only basis for choosing therapy, and adequate clinical results are often attained with less potent drugs. Also, *Bacteroides fragilis* is mentioned as a "worst case"; as noted, it is not the most common *Bacteroides* species in gynecologic infections and other species are often more susceptible. As a class, cephalosporins have little or no activity against the enterococci;

consequently cefoxitin monotherapy misses the enterococci. Cefoperazone, moxalactam, cefotaxime and imipenem-cilistatin have also been used as monotherapy in pelvic infections. These agents typically have activity against gram-negative facultative bacteria and some anaerobes, and although the susceptibility of *Bacteroides fragilis* is limited, most other *Bacteroides* species are inhibited. Imipenem, however, provides coverage for enterococci and *Bacteroides fragilis*. Despite this, its role in treatment of postoperative sepsis has been limited, and the extent to which it will be used in the future is uncertain.

Another therapeutic option in cases of postoperative infection is to cover anaerobic bacteria with metronidazole and to use gentamicin to provide coverage of the gram-negative facultative rods. This combination has the disadvantage of somewhat weak activity against gram-positive organisms. Metronidazole displays excellent activity against anaerobic gram-negative bacteria, but it is less potent with respect to anaerobic gram-positive cocci. The combination of metronidazole and gentamicin is also weak against the enterococci, *Staphylococcus aureus*, and group B streptococcus. Similar coverage can be achieved by substituting trimethoprim-sulfamethoxazole for the aminoglycoside, although this combination requires clinical testing to ensure that it furnishes adequate clinical results.

It is notable that clinical experience indicates that the various antimicrobial regimens discussed give approximately comparable results. This is not surprising if one considers the nature of postoperative infections. Because these infections are polymicrobial, the interactions between the various bacterial species probably play a major role in the infectious process. One of the effects of antibiotics may be to disrupt synergistic interspecies relationships; as a consequence, it may not be necessary for therapy to inhibit every organism at the infected site.

Postpartum Endometritis and Infection After Cesarean Delivery

The microorganisms that cause these infections are largely the same as those involved in postoperative infection. Hence, the antibiotic choices are similar to those employed in postoperative infections, with a few exceptions. Because the antecedent to many cases of postpartum infection is chorioamnionitis, which frequently involves group B b-hemolytic streptococci, the inclusion of this organism in the spectrum of activity of the chosen therapy is appropriate. Some evidence exists to suggest an occasional role for mycoplasmas in postpartum infection, but the need to include coverage for these organisms has not been established.

Because chorioamnionitis is frequently involved in the genesis of postpartum sepsis, one may consider antibiotic use in women with chorioamnionitis. However, the clinical result of antibiotic treatment for amniotic infection is blunting of the maternal symptoms, but little benefit to the fetus. In fact, it was considered preferable in the past to avoid treating the infant with antibiotics in utero because signs of infection could be masked when the baby is born. Some investigators have questioned this prohibition and have indicated that no direct adverse effect on the infant is demonstrable. The actual protocol followed may require input from the pediatric staff because many pediatricians consider the baby who received antibiotics in utero as committed to postpartum antibiotic therapy, whereas other pediatricians may not consider this to be a problem. However, one should not assume that antibiotic therapy reaching the fetus from the maternal circulation will prevent all infectious complications resulting from chorioamnionitis because the perfusion of the fetal lungs, which may be the primary site of infection in utero, is limited until after birth. In cases of chorioamnionitis, the infant can be delivered and treated separately from his or her mother even if intrapartum antibiotics are employed. The maternal infection can then be treated as noted for postpartum sepsis.

Infection Complicating Malignancy

Infections may complicate the postoperative course in women with gynecologic malignancy. The organisms of primary concern are the same ones of interest in other postoperative gynecologic infections. However, both *Pseudomonas aeruginosa* and *Clostridium perfringens* are more commonly encountered in the flora of patients with gynecologic malignancy. In addition, because the malignancy may be associated with some degree of immunocompromise, a bactericidal antibiotic may be most desirable. Although not specifically addressing gynecologic malignancy, clinical trials involving the use of cefoperazone indicated that it provided good activity comparable to that provided by combination therapy. This drug has a bactericidal mechanism of action and is active against *Pseudomonas aeruginosa*.

Bacteremias

Bacteremias arising as a complication of intraabdominal infections may be suspected of involving *Escherichia coli* or *Bacteroides* sp. and may be initially treated with cefoxitin or an aminoglycoside-clindamycin combination until more specific culture information becomes available; during pregnancy the combination of ampicillin plus gentamicin is appropriate. Bacteremic disease in the pregnant patient may be due to listeriosis. The drug therapy of choice is ampicillin, which may be given by mouth in most cases because the patient usually is not too ill to take oral medication. Erythromycin is an alternative in case of penicillin hypersensitivity.

Bacteriuria

Uncomplicated urinary-tract infection is usually treated with a 3-day course of trimethoprim-sulfamethoxazole (160 mg and 800 mg, respectively, twice daily [bid]), trimethoprim alone (200 mg bid), ciprofloxacin (250 mg bid), norfloxacin (400 mg bid), or nitrofurantoin (100 mg four times daily [qid]). The quinolone antimicrobial agents are not used in the pregnant patient. Recurrent infection should be treated for 7-10 days with follow-up examination after therapy has been completed.

If pyelonephritis occurs in pregnancy and is due to gram-negative bacteria, ampicillin or amoxicillin with or without clavulanic acid is a reasonable therapeutic choice; an alternative to ampicillin is a cephalosporin such as cefazolin. Therapy should be continued for 14 days, and the patient's condition should be carefully monitored.

Sexually Transmitted Diseases

Because of the public health significance of sexually transmitted diseases, the Centers for Disease Control (CDC) have become involved not only with bringing some degree of uniformity to the reporting of these conditions but also with establishing guidelines for therapy. The therapeutic guidelines represent the minimum standard for treatment. Only a brief summary of the antibiotics used to treat these conditions is presented here, and the reader is referred to the CDC publications for the complete description of the standards.

SYPHILIS

Treponema pallidum is exquisitely sensitive to penicillin and has not demonstrated any propensity toward the development of resistance. The main consideration in therapy is the use of a form of penicillin that provides the pharmacologic properties that are most suitable for treatment of this disease. Consequently, the specific dosage form depends on the stage of the disease.

In adults, early disease is treated with a single dose of benzathine penicillin G (2.4 megaunits), and late syphilis is treated with three 2.4-megaunit doses at weekly intervals. The same therapy is applied to pregnant patients. In allergic patients the alternative therapy is a 15-day course of tetracycline 500 mg qid or doxycycline 100 mg bid for early disease and a 30-day course for late disease. Tetracycline is contraindicated in pregnancy, and erythromycin (500 mg by mouth qid) is the approved alternative. Erythromycin is to be given for 15 days in cases of early syphilis and 30 days in cases of late syphilis. Obviously, patient compliance will be problematic, and syphilis in pregnancy is too important a problem to risk haphazard therapy. Among the unapproved alternatives, ceftriaxone appears to be a suitable drug, and because administration is by the intramuscular route, patient compliance is less of a problem. Another alternative is to desensitize the allergic patient and proceed with penicillin therapy.

For neurosyphilis the approved therapy consists of a 10- to 14-day intravenous course of aqueous crystalline penicillin (2-4 megaunits) given every 4 hours until a total of 12–24 megaunits has been administered. This is followed by 3 weekly doses of 2.4 megaunits of benzathine penicillin G. Congenitally infected infants with abnormal spinal fluid findings are treated with a 10-day course of aqueous crystalline penicillin G (50,000 U/kg divided between two injections). The importance of adequate follow-up in cases of syphilis, especially those that occur in infants, cannot be overemphasized.

GONORRHEA

Therapy for *Neisseria gonorrhoeae* has undergone recent changes that have been necessitated by the emergence of resistance problems, including penicillinase production, tetracycline resistance, and chromosomally mediated resistance. In addition, the frequent presence of *Chlamydia trachomatis* has prompted coverage for this organism in the course of gonococcal therapy. In uncomplicated gonococcal infection of the cervix, the patient is treated with ceftriaxone (250 mg intramuscularly), followed by a 7-day course of oral doxycycline (100 mg bid). If penicillin is contraindicated, the alternative is to treat the patient with spectinomycin, followed by 7 days of tetracycline. Other alternatives include norfloxacin, ciprofloxacin, cefuroxime axetil plus probenecid, cefotaxime, or ceftizoxime. For specific dosages the CDC guidelines should be consulted.

CHLAMYDIA

Chlamydia trachomatis infection may be manifest as mucopurulent cervicitis or may be involved as part of an upper-genital-tract infection. Doxycycline is the drug of choice to treat chlamydia and is often given on suspicion of infection because cultures do not always yield the organism. Therapy for cervicitis is a 7-day course of either tetracycline (500 mg by mouth qid) or doxycycline (100 mg by mouth bid). An alternative to tetracycline is 7 days of erythromycin, which is also appropriate for use in pregnancy, although its pharmacologic properties are less than ideal. When erythromycin is not tolerated in the nongravid patient, sulfisoxazole (500 mg by mouth qid for 10 days) may be used. A detailed description of the therapeutic regimens can be found in the CDC guidelines.

PELVIC INFLAMMATORY DISEASE

Current therapy for pelvic inflammatory disease is based on the recognition that lower-tract organisms, in addition to the sexually transmitted agents *Neisseria gonorrhoeae* and *Chlamydia trachomatis,* are involved in acute salpingitis. No ideal therapy has been discovered, and several options are available. Doxycycline, which is given as part of most regimens for coverage of the gonococcus and chlamydia, is continued for a course of 10-14 days. Inpatient therapy may be a combination of cefoxitin and doxycycline, and outpatient therapy may be immediate intramuscular doses of cefoxitin and probenecid followed by 14 days of doxycycline or tetracycline. Alternative inpatient therapy may include clindamycin plus gentamicin. After discharge from hospital, doxycycline is continued for 10-14 days. For patients who do not tolerate doxycycline, an alternative form of outpatient therapy is erythromycin (500 mg by mouth qid for 10-14 days). Another option for outpatient therapy is oral trimethoprim-sulfamethoxazole plus oral clindamycin therapy, both continued for 10-14 days.

HAEMOPHILUS DUCREYI

The treatment of chancroid is best accomplished with a single intramuscular injection of 250 mg of ceftriaxone, a 7-day course of erythromycin base (500 mg by mouth qid), or a 3-day course of ciprofloxacin (500 mg by mouth three times daily) unless the patient is pregnant or less than 16 years of age. The older therapy involving trimethoprim-sulfamethoxazole is becoming obsolete due to antimicrobial resistance, but amoxicillin plus clavulanic acid (7-day course) has been effective in trials conducted outside the United States.

Viral Infections

The treatment of patients who have any of the viral infections described in chapter 10 involves more than just the administration of an antiviral agent. However, in the following paragraphs a brief description of the available antiviral therapy is presented.

HERPES SIMPLEX VIRUS

The treatment for genital herpes infection is primarily acyclovir, but in cases in which the ulcers show evidence of secondary infection, an antimicrobial agent may be considered as adjunctive therapy. Acyclovir is available as either an oral medication or as a topical preparation. Oral therapy for the first occurrence of genital herpes employs acyclovir (200 mg by mouth five times daily) given for 7-10 days. Prolonged oral therapy for multiple symptomatic recurrences is usually provided to immunocompromised hosts.

HUMAN IMMUNE DEFICIENCY VIRUS

The patient with acquired immune deficiency syndrome (AIDS) will have a variety of infectious complications that will need to be addressed as specifically as possible. The use of the antiviral agent azidothymidine, now known as zidovudine, is a means for slowing the course of the disease but not curing it. Patients who have symptoms associated with human immune deficiency virus (HIV) infection, either AIDS or a T-helper cell concentration of <500/mL, are candidates for therapy. Individuals who show serologic evidence of having been infected with HIV but who do not have symptoms are not candidates for therapy. The drug is administered orally every 4 hours in 100-mg doses. Currently this is the only approved treatment, although additional drugs will probably become available in the future.

Although the use of zidovudine in pregnancy has been controversial in the past, many investigators have used it because of seriousness of AIDS infection in pregnant women. Fairly careful monitoring of these nonapproved uses has indicated that the drug appears to have a reasonable risk-to-benefit relationship in pregnancy.

HUMAN PAPILLOMAVIRUS

No specific antiviral drug is available to treat human papillomavirus, and genital warts are treated topically with either podophyllin or 5-fluorouracil. Both of these compounds are toxic and should not be used without a thorough understanding of their proper application. Other local therapy includes liquid nitrogen cryotherapy or laser ablation.

13
Chemoprophylaxis

Antibiotics have long been used to prevent infections, but numerous changes have occurred in the way antimicrobial agents are administered to achieve prophylaxis. Among the important contributions to antibiotic prophylaxis have been studies suggesting how prophylaxis may be used most effectively, but how prophylaxis actually works is not completely understood. Many aspects of antimicrobial prophylaxis as it applies to the obstetric and gynecologic specialty remain primarily limited to clinical observations comparing one prophylactic regimen with another, with only a few studies scientifically addressing the mechanisms of prophylactic action. This chapter will present the basic principles of antimicrobial prophylaxis that have grown partly from scientific investigation but mostly from an attempt to consider rationally the way in which these drugs may be employed. Such principles must be seen as being subject to reconsideration as a better scientific foundation for them is developed.

BASIS OF ANTIMICROBIAL PROPHYLAXIS

Principle 1: Timing of Drug Administration

TENET 1: THE DRUG IS GIVEN AS NEAR TO THE TIME OF CONTAMINATION AS POSSIBLE

The pioneering work of Burke has been interpreted as indicating that prophylaxis represents a window of opportunity in which the antibiotic must be given before the introduction of bacteria; an antibiotic given after organisms were introduced into tissue was less effective as the time increased from infection to antibiotic administration. If the antibiotic was not given until 3 hours after infection was initiated, no protective value was recorded. These investigations employed staphylococci implanted into the skin of laboratory animals, and prophylaxis consisted of intravenously administered penicillin. The effectiveness of prophylaxis was based on the size of the lesion. This study may not be absolutely applicable to all types of prophylactic antibiotic use. In obstetric and gynecologic infections, the disease process frequently involves multiple species of microorganisms that may act in concert to produce symptoms. Soft tissues within the abdomen are usually involved, in contrast to the skin lesions studied by Burke. Although the laboratory studies have been valuable in providing some basis for prophylaxis, they also point out the need for controlled studies that address the use of prophylaxis in animal models that more closely resemble the types of infection seen in obstetric and gynecologic patients.

The above-mentioned observation by Burke has become recognized as a key principle in antimicrobial prophylaxis and has established the first tenet of antimicrobial prophylaxis, namely, that tissue levels of antibiotic should be present when the incision is made. This tenet is based on the belief that the infection is initiated with the contamination of the surgical wound.

As noted in earlier chapters, we do not understand how the organisms that contaminate the surgical wound actually cause infection. The simple fact that an organism such as *Bacteroides fragilis*, *Escherichia coli*, or even *Clostridium perfringens* contaminates the wound is clearly not sufficient to incriminate the organism as an etiologic agent. The organisms apparently must proliferate and cause damage. Therefore, by inference it is believed that antimicrobial prophylaxis does not function by preventing wound contamination.

Several studies that evaluated the bacterial flora of the vaginal vault before and after hysterectomy in patients who received prophylactic antibiotic or placebo indicated that even in the presence of antibiotic, an abundant flora was present in the vaginal vault. In addition, the surgical procedure caused an increase in the diversity of the flora. These studies were not quantitative and thus did not reveal whether there was also an increase in the population density of the vaginal microorganisms. In those individuals who received antibiotic prophylaxis, an increase in the diversity of the flora also occurred, although it appeared to be blunted by the prophylactic antibiotic. Thus, the way in which antimicrobial prophylaxis functions in gynecologic surgery may be to alter the magnitude of the postsurgical proliferation of the flora, but only when the drug is given early with respect to the contamination of the tissues.

Because it is recognized that the organisms that potentially infect the surgical site are not eliminated by the prophylactic drug, it may be speculated that one of the specific effects of the antibiotic is to prevent the organisms from rapidly adapting to and exploiting the environment of the operative wound. This may possibly be accomplished by preventing the organisms from fully expressing virulence attributes. Studies of the effects of very low concentrations of antibiotics have demonstrated that at concentrations that do not completely prevent bacterial growth (sub-minimum inhibitory concentrations), inhibition of the synthesis of certain virulence attributes may occur.

TENET 2: THE DURATION OF PROPHYLAXIS IS LIMITED TO THREE DOSES OF THE ANTIMICROBIAL AGENT

Although the animal studies mentioned have been interpreted as indicating that prophylactic antibiotics must be present before bacterial contamination occurs, they also show that some prophylactic efficacy persists for 1 or 2 hours after initiation of infection. By 3 hours after infection, the efficacy of prophylaxis is very limited. There are clinical situations in which prophylactic antimicrobial use is contemplated, but it is not possible to administer drug before contamination. For example, in prophylaxis of neonatal ophthalmia, the contamination occurs at birth, but the eye drops or ointment is administered later. In providing prophylaxis against postpartum sepsis in high-risk women undergoing cesarean delivery, many physicians prefer to administer the antibiotic after the cord has been clamped to prevent transfer of antibiotic to the fetus. In such a case, the antibiotic will not reach the wound until after bacteria have contaminated the operative site.

The observation that the efficacy of prophylaxis diminishes in proportion to the length of time between wound contamination and antibiotic administration has in part been responsible for another innovation in the way antibiotic prophylaxis is used clinically. Early attempts to reduce morbidity began in the 1940s when sulfonamides represented essentially all available antimicrobial agents. Many early studies made use of prolonged courses of drug, in some cases as much as a week, to ensure that the full benefit of the drug would be realized. If the studies that Burke performed in his animal model are applicable to prophylaxis of surgical infections in humans, it may be concluded that any benefit that accrues to patients given antibiotics beyond 3 hours after surgery is due to the therapeutic rather than the prophylactic effect of the drug. Therefore, the history of antimicrobial prophylaxis has seen drug administration reduced from days to no more than three doses, with the first dose given before beginning surgery. More recently, single-dose prophylaxis has been employed. It should theoretically be as effective as three doses given over an 8- to 12-hour period because the second and third doses actually fall outside the realm of prophylaxis in terms of timing.

TENET 3: THE DRUG USED FOR PROPHYLAXIS SHOULD HAVE A RELATIVELY LONG HALF-LIFE

This is an intuitive recommendation that may seem to contradict the aforementioned concept that the critical period in initiation of infection and interference with pathogenesis is very early, perhaps in the first hour after contamination. Nevertheless, there is logic behind this concept. First, as a practical matter, it is important to have the highest levels of antibiotic possible in the serum and tissues at the time the surgical wound is created. If antibiotic is administered on call to the operating room, delays in commencing the surgery may mean the infusion is completed and the serum concentration is on the decline before the operation begins. Second, it is not known at what time the prophylactic antibiotic exerts is action. On the basis of the previously described animal study, host-drug-microorganism interactions that determine whether postoperative sepsis will occur seem to happen at the moment of contamination. However, intuition indicates that these interactions will probably continue for some time after the surgical site has been contaminated. Hence, a long-lasting antimicrobial agent has a greater ability to influence the host-microbe interaction beyond the initial few minutes after the tissues are contaminated. A drug with a reasonably long half-life will sustain serum levels in those cases in which multiple doses of prophylactic antibiotic are used. There is currently no absolute proof that an antibiotic with a long half-life must be used or even should be used, but intuition dictates that if all other factors are equal in selection of a prophylactic regimen, the drug with the longer half-life will be a better choice.

Principle 2: Selection of Antimicrobial Regimen

TENET 1: THE SPECTRUM OF ACTIVITY OF AN ANTIBIOTIC HAS RELEVANCE FOR THE ORGANISMS LIKELY TO CAUSE THE INFECTION

It is a logical assumption that effective antimicrobial prophylaxis should employ a drug that inhibits the organism or organisms that are likely to cause infection. This concept proves to be sound for prevention of infections due to a single, well-defined infectious agent. However, most uses of prophylaxis in obstetrics and gynecology are intended to prevent polymicrobial infections arising from members of the normal genital microflora that typically contaminate surgical sites. In these types of infections it is impossible to identify the organisms responsible for the symptoms, and it is equally impossible to determine which organisms are acting synergistically to produce disease. It might be assumed that adequate prophylaxis would require coverage for all microorganisms present in the operative site. However, in dozens of clinical studies, adequate and comparable results have been attained with antibiotic regimens that have different antibacterial spectra. The reason for these comparable results is not known. However, it seems that if polymicrobial infections are characterized by intricate interspecies interactions, an antibiotic that does not inhibit all species in a polymicrobial infection may nevertheless disrupt the delicate ecology of infection, allowing the host defenses to prevail.

The spectrum of antimicrobial activity of drugs used for prophylaxis should be an important consideration. Although it is not necessary to provide coverage for all organisms at the operative site, it is reasonable to provide inhibitory activity against some of the major organisms. Thus, for prevention of postoperative sepsis in most surgical procedures involving the female genital tract, coverage for gram-negative rods and some anaerobic coverage is effective.

The inclusion of anaerobic coverage in the antimicrobial spectrum of the prophylactic drug is an important issue. One of the special attributes of certain of the *Bacteroides* species is their ability to produce abscesses. It is not clear if the dynamics of abscess production are independent from the more acute soft-tissue infection that may occur earlier in the course of intraabdominal infection. If abscesses arise independently, then prophylaxis that does not include antianaerobic activity may suppress acute signs of infection but allow abscess formation in a few patients. Some investigators have indicated that this is indeed the case. Nevertheless, most abscesses arise in an environment of mixed infection, and independent abscess production by anaerobic organisms as a process separate from the cooperative effects of aerobic or facultative organisms is probably uncommon.

In summary, effective prophylaxis should include coverage for the major organisms that are anticipated in postoperative infections, but comprehensive coverage for all organisms is not required. In preventing monoetiologic infections, the antimicrobial drug should show efficacy against the organism of concern.

TENET 2: THE ANTIBIOTIC DOES NOT HAVE A UNIQUE ROLE AS A THERAPEUTIC AGENT

Not all antimicrobial agents should be considered for prophylactic use. Certain antibiotics have proven to be far too important in specific situations to be used for prophylaxis. Among the antibiotics frequently used prophylactically in obstetric and gynecologic surgery are the first- and second-generation cephalosporins, and to a lesser extent the extended-spectrum penicillins, because they have the desired spectrum of activity but are not dedicated to special uses. This concept is based on the concern that each use of an antibiotic induces selective pressures to increase antimicrobial resistance. The more a particular antimicrobial agent is used, the less confident one can be about the probable effectiveness of the drug in a clinical situation. However, the degree to which antibiotic prophylaxis creates altered resistance patterns is unknown. This uncertainty has caused some investigators to question whether the prohibition against the use of some antibiotics for prophylaxis is warranted. Despite this attitude, as new and highly potent agents are developed, it seems rational to use them to their best therapeutic advantage as long as the older, general-purpose antibiotics are still providing effective prevention of postoperative infection.

A concept related to that of holding therapeutic agents in reserve is the belief that when prophylaxis fails and the patient shows evidence of infection, the same antibiotic used for prophylaxis should not be used for therapy. Consequently, if a particular antibiotic is favored as a therapeutic agent, it is no longer an option for use as a prophylactic agent.

TENET 3: THE ANTIBIOTIC SHOULD HAVE AN EXCEPTIONALLY GOOD SAFETY PROFILE

It must be conceded that most patients who receive a prophylactic antibiotic are not at grave risk of developing a serious infection. Although prophylactic antibiotics are generally used when patients undergo procedures associated with substantial infectious morbidity, most morbidity recorded is limited to transient temperature elevations. In such cases, the primary value of prophylaxis is that it prevents minor morbidity and the associated cost of extra hospital days, infection work-up, cultures, and therapeutic antibiotics. In view of this fact, the use of a highly toxic antibiotic to prevent what is frequently a trivial infection cannot be justified.

TENET 4: THE ANTIBIOTIC SHOULD BE COST-EFFECTIVE

This concept is related to the previous tenet regarding the safety of the drug. Because many instances of morbidity are not serious, neither toxic nor expensive antibiotics should be used. Generally, antimicrobial prophylaxis is cost-effective because even trivial morbidity is expensive, and the cost of serious morbidity is likewise substantial. However, adequate prophylactic effects may usually be obtained by use of inexpensive drugs rather than the newest (and hence most expensive) agents with extremely broad spectrums of activity. To use an expensive drug when a less expensive alternative works as well is tantamount to a waste of resources. Because prophylaxis is applied to all individuals (ie, all women undergoing vaginal hysterectomy) rather than to a selected group, as in the case of therapy (eg, only those who develop signs of infection), waste related to use of unduly expensive drugs for prophylaxis is systematically compounded. Most hospital formularies are now designed to enforce cost consciousness and will discourage use of expensive drugs for prophylaxis.

Principle 3: Selection of Patients to Receive Prophylaxis

TENET 1: THE PATIENT HAS A SIGNIFICANT RISK OF INFECTION

Both from the literature and from the local experience of the physician it is possible to gauge the magnitude of the problem of infectious morbidity after specific surgical procedures. There is no magic formula for deciding what rate of infection is acceptable before intervening by providing antimicrobial prophylaxis. The decision as to whether prophylaxis is warranted is based on the consensus of specialists in the field who set the standard of care, and by evaluation of the infection rate and the cost of those infections in the local setting. The obvious pitfall that must be avoided is the overly enthusiastic use of prophylaxis on the grounds that it cannot hurt anything. Antibiotics tend to be among the safest of drugs available. Yet adverse reactions can result from their use. In addition, the local and global patterns of antibiotic resistance may be influenced by inappropriately liberal use of antibiotics.

TENET 2: THE INFECTION TO BE PREVENTED IS NOT TRIVIAL

As stated above, much of the febrile morbidity that occurs in patients undergoing gynecologic operations is technical morbidity that will resolve spontaneously without therapy and without sequelae. Much of this morbidity is prevented by antimicrobial prophylaxis. But the real benefit, which should be at the heart of prophylactic antibiotic use, is prevention of serious morbidity. The physician should be cognizant of the prevalence of serious infectious complications of surgical procedures both in his or her own practice and in the literature. In judging the value of prophylactic regimens that are recounted in numerous journal articles, one must take note of the contribution of serious infections to the overall prevalence of morbidity presented in these studies.

TENET 3: THE PROPHYLACTIC REGIMEN HAS BEEN DEMONSTRATED TO HAVE EFFICACY

Numerous antibiotic regimens have been evaluated for their effectiveness, and the results are available in the obstetric and gynecologic literature as well as the surgical literature. The studies attempting to demonstrate prophylactic efficacy are largely comparative studies, and some have employed placebo as a control. In recent years, it has become so firmly established that prophylaxis prevents infections that most institutional review boards consider it unethical to deny patients prophylaxis by placing them into placebo control groups. Consequently, most studies compare the postoperative infection rates in groups of women treated with different antibiotics. The validity of such comparisons is based on the comparability of the two patient groups and the lack of bias in conducting the study and evaluating the results. The comparability of the two groups is enhanced by having adequate numbers of patients and assigning them to a particular drug regime on a statistically random basis. Bias is obviated by blinding the study. A double-blind study is one in which neither the physician not the patient knows which drug was administered. In this way the evaluation of the patient's postoperative course is not clouded by the physician's expectations. Even when studies are conducted in this manner, they cannot provide all the information that is desirable. Perfect matching of the two groups is never possible. Comparisons between studies are not valid, and one must resist the temptation to compare the efficacy of one antibiotic observed in one study to the efficacy of a second antibiotic in a second study.

It can be stated from the dozens of available studies of prophylactic antibiotics that prophylactic antibiotics reduce infectious morbidity after surgery and that no clearly superior prophylactic regimen has emerged from this volume of work.

PRACTICE OF ANTIMICROBIAL PROPHYLAXIS

The specific clinical situations in which prophylactic antibiotics are used are continually being reevaluated. The fact that there is a precedent for prophylaxis should not necessarily be construed as meaning such use is appropriate. The long-standing indications for prophylactic antibiotic use are still considered to be appropriate standards of care, namely, ophthalmic prophylaxis to prevent neonatal ophthalmia and preventive antibiotic for individuals with rheumatic heart disease who are to undergo surgical procedures.

In gynecologic surgery, the use of prophylaxis in vaginal hysterectomy is well established because of the high morbidity rates seen postoperatively. Prophylaxis in abdominal hysterectomy has not always been advocated because most surveys of infectious morbidity after abdominal hysterectomy found an intrinsically lower infection rate than that associated with vaginal hysterectomy; however, antibiotic prophylaxis is now considered appropriate management. Undoubtedly radical procedures and surgical procedures on individuals with significantly compromised host defenses are considered settings in which prophylaxis is exceedingly beneficial. When in the course of a surgical procedure the bowel or bladder is accidentally entered, the damage is repaired and antibiotics are given to preempt the development of infection.

More controversial is the use of antibiotic prophylaxis in patients undergoing tubal microsurgery. The rationale for its use is the possibility that infection could seriously compromise the delicate repairs to the fallopian tubes. Extensive laser treatment of large areas of the vagina is associated with a high rate of occurrence of a severe febrile reaction. This raises the question of whether such patients would be well served by antibiotic prophylaxis.

In obstetric practice, emergency cesarean delivery is accompanied by a significant likelihood of postpartum infection. The risk of infection is further increased by the presence of certain demographic and other predisposing factors that help the physician select patients who may require prophylactic antibiotics. As noted previously, it is common practice to withhold the antibiotic until after the cord has been clamped. This prevents exposure of the infant to the antibiotic, which may not be of benefit and could serve to mask the signs of infection.

In the case of exposure to certain sexually transmitted diseases, antimicrobial prophylaxis may be administered, although it may be argued that this represents early therapy.

In summary, the physician should realize that prophylaxis provides an opportunity to avert both the clinical problems of infection in the patient and the discomfort and disruption of the patient's convalescence that accompany postoperative infection. As with many advances in medicine, there are pitfalls that can to some extent be avoided by adherence to the principles set forth in this chapter. The use of antibiotics as a preventative measure should be as rational an undertaking as use of antibiotics for therapeutic purposes.

14

Immunoprophylaxis

Overview
Available Vaccines

- Cholera
- Diphtheria
- *Haemophilus influenzae*
- Hepatitis A
- Hepatitis B
- Influenza
- Measles
- Mumps
- *Neisseria meningitidis*
- Plague
- Poliomyelitis
- Rabies
- Rubella
- *Streptococcus pneumoniae*
- Tetanus
- Tuberculosis
- Tularemia
- Varicella
- Yellow Fever

OVERVIEW

The immune response is an essential part of preventing microorganisms prevalent in the environment from overwhelming the body, as evidenced by the life-threatening failures of the immune system that are exemplified by severe combined immune deficiency syndrome or acquired immune deficiency syndrome. The ability of the host to rally its defenses against microbial invaders is certainly important to our human survival, but an aspect of the immune response that may be of equal importance is the anamnestic response. Some of the most virulent pathogens are capable of killing the host before an immune response is able to develop and alter the course of the disease.

Since the first immunization accomplished by Jenner in 1796, exploitation of the secondary or anamnestic immune response has paralleled the development of knowledge about the primary immune response. Indeed, the main role of immunization is to elicit a primary immune response that primes the host for a secondary response when the virulent organism is encountered naturally. Vaccinations that cause the host to elicit its own immune reaction are referred to as active immunizations. The bacteria, viruses, fungi, and protozoans that can infect the human host number in the hundreds. Nevertheless, vaccination is not practiced against all of these infectious diseases. Certainly, for many of these agents there is no effective vaccine. But for those diseases for which a vaccine has been developed, the impact on morbidity and mortality has probably been a part of shaping history.

Although the strategy of most vaccinations is to administer an antigen to an immunocompetent host to elicit an immune response, there are some circumstances in which the administration of preformed antibodies is required. This method is termed passive immunization because it involves antibodies not produced in the host that is being immunized. Such vaccinations are employed in cases where the disease may develop before active immunization is able to produce effective concentrations of antibodies. Another use of preformed antibodies in the form of hyperimmune serum is to attenuate the potential untoward effects of live-virus vaccines in some individuals. Finally, hyperimmune serum may be required in the event that no active immunizing material has been developed.

There are two general forms of vaccine materials used for active immunization. One contains living, but attenuated, organisms, and the other consists of killed organisms or some nonliving isolated component of the microorganism. The whole subject of vaccination is usually one for immunologists, internists, and pediatricians, because many immunizations are administered during childhood. However, there are three situations in which the obstetrician is concerned about immunization. First, the obstetrician is concerned that women who become pregnant have been properly vaccinated against infectious diseases that may severely affect the outcome of pregnancy and that can be prevented by vaccination. Second, the obstetrician has concern for exposure of the pregnant patient to live viral vaccines during pregnancy because it is not clear whether some of these may cause harm to the fetus. Third, the transplacental passage of antibodies, which may be protective, is of interest as well.

Far fewer bacterial infections than viral infections are amenable to prevention through vaccination. Bacteria tend to be antigenically more complicated and subject to great variation in antigenic composition. A single bacterial species may exist as many serotypes and biotypes and depend on a variety of virulence factors and physiologic properties to cause symptoms in the host. Viruses, in contrast, frequently depend on only a few surface components to engage in the critical interactions with host cells that ultimately result in infection. Consequently, the production of antibodies that specifically neutralize the infectivity of viruses may be a less complicated task than finding an antigenic determinant that protects against bacterial infection.

Vaccination against viral diseases is of special significance because the availability of antiviral drugs is limited, and there are few drugs that may be used prophylactically to prevent viral infections. However, with respect to bacterial infections, chemotherapy and chemoprophylaxis are well developed. Routine vaccination against bacterial disease is practiced in the case of pertussis, diphtheria, and tetanus. The latter two depend on the use of toxoids as immunogens because the entire disease processes can be averted by neutralizing the toxins of either tetanus or diphtheria. The whole bacterial vaccine used to prevent pertussis is an exception with respect to its antigenic complexity and the greater propensity to cause untoward effects compared with many other vaccines.

The current state of the science of immunoprophylaxis permits the prevention of a very large number of potential cases of nine childhood diseases, some of which have significant implications for the well-being of the pregnant woman or her fetus. Unfortunately, there are few tests available to routinely evaluate the effectiveness of vaccine protection. The presence of antibody titers against rubella can be tested clinically, and because of the devastating effects of the congenital rubella syndrome, this test is appropriate for women seeking obstetric care. For other viral disease that can be prevented by vaccination, one must rely on the patient's history. Has she been vaccinated, or has she had the viral illness? Routine vaccinations in childhood include diphtheria, pertussis, tetanus, measles, mumps, rubella, and polio, although immunity wanes with time and diphtheria and tetanus susceptibility should be assumed if the last booster was given more than 10 years previously. In addition, a significant percentage of women will also have immunity to varicella because of having had the disease during childhood.

In the future, additional vaccines will become available, and some of these will have a direct impact on infections in pregnancy. Vaccines based on gonococcal pili and the polysaccharide of the group B streptococcus, as well as an acellular pertussis vaccine, are anticipated within the next decade. Viral diseases for which new or improved vaccines may become available in the next 10 years include cytomegalovirus, varicella virus, hepatitis A and B, and herpes simplex types 1 and 2, as well as others of less interest to obstetricians. Vaccine development is a complex and arduous process, and testing for safety and efficacy poses vast problems. Consequently, the National Academy of Sciences Institute of Medicine predicts that vaccines for *Chlamydia trachomatis*, *Treponema pallidum*, and human immune deficiency virus will not be available in the next decade.

In the remainder of this chapter a summary of the available immunization materials will be presented. At the outset, the following generalizations should be noted: Vaccination should be avoided during pregnancy, but if it is deemed necessary, it should involve only nonviable vaccines. Live attenuated vaccines should be avoided for 3 months before conception, as well as during pregnancy. There is a theoretical possibility that the vaccine virus may cross the placenta and do damage to the fetus, although untoward effects on the fetus have not been documented. In the cases of rabies and yellow fever, the mortality is so high that the benefit of vaccination in exposed individuals is considered to outweigh the potential risk. All cases of pregnant women receiving live attenuated vaccines or women receiving the vaccines within 3 months of conception should be reported to the Centers for Disease Control (CDC). Local and state health departments can assist in getting the appropriate information to the CDC.

AVAILABLE VACCINES

Cholera

Cholera is acquired by ingestion of infected, contaminated water, resulting in attachment, replication, and production of toxin by the organism in the gut lumen. The disease is ordinarily limited to endemic areas including the Bengal area of India and Bangladesh, but during epidemics it has spread to other parts of the world, including southeast Asia, Africa, and western Europe. During pregnancy the organism may cause severe disease, especially during the third trimester, when loss of the fetus may occur.

Cholera is primarily related to inadequate sanitation, and in endemic areas boiling water supplies for at least 10 minutes can be a means of preventing disease. Eating uncooked or undercooked shellfish has been the cause of rare outbreaks in the United States. In endemic areas unwashed vegetables should not be eaten, and only bottled or boiled water should be ingested.

Vaccination has tended to depend on low-efficiency products, including killed bacterial vaccines or live attenuated organisms. In any case, the vaccine must be given orally to induce intestinal antibody. Apparently the most promising protection comes from genetically altered strains that have defective toxin genes as well as bacterial antigens. Vaccines are not generally available, and for pregnant women, protection is best achieved by either postponing travel to endemic areas or by attention to proper preparation of all materials to be ingested.

Diphtheria

Diphtheria is caused by a single factor, the diphtheria toxin. This exotoxin can readily be transformed into a toxoid by Formalin treatment and provides an effective immunogen. This toxoid may be given during pregnancy, and it may be provided as part of routine antenatal care to patients who have not had their tetanus and diphtheria immunity boosted within 10 years. The mortality due to this disease is significant, and the prevalence of susceptibility may be quite high in some population groups, especially older adults who have not had their immune status updated. The immunoglobulin G (IgG) antibodies produced by the mother cross the placenta, and along with anti-tetanus antibodies (tetanus and diphtheria toxoids are usually coadministered) they protect the neonate during the first 6 months of life. One of the important reasons for updating the immunity of the pregnant woman is to provide her newborn with the transplacental protection afforded by her antibodies.

Passive immunization is available for individuals with a presumptive diagnosis of diphtheria. The earlier in the course of the disease the antitoxin is administered, the better the patient's prognosis is. Host cells already intoxicated will not be affected by antitoxin; rather, antitoxin serves to prevent further binding of unbound toxin to unintoxicated cells. The antitoxin is of equine origin, so hypersensitivity to horse serum must be evaluated before administering antitoxin, because anaphylaxis is a potential hazard. Passive immunization should also be accompanied by active immunization.

Haemophilus influenzae

This organism is a respiratory pathogen that is a significant cause of morbidity and mortality in young children below the age of 6 years. It is also the leading cause of meningitis in this age group. The serotype associated with most cases of illness is type B. Consequently, the immunogen consists of type B polysaccharide capsule material conjugated to diphtheria toxoid to enhance its immunogenicity. Polysaccharide antigens alone are typically less immunogenic than protein antigens, hence, the conjugation of a protein to the *Haemophilus* antigen. It should be noted that the toxoid present as part of this vaccine material is not considered an adequate means of vaccinating against diphtheria.

Although rare reports of *Haemophilus influenzae* as a cause of intraamniotic infection have appeared in the literature, there is no information regarding prevention of this condition by immunization against *Haemophilus influenzae* type B. Currently the condition is sufficiently uncommon that it seems unlikely to require an evaluation of vaccination strategies. There are no known fetal complications related to maternal infection. Adults appear to respond to exposure to the organism with the production of antibodies both to the capsular antigen and to the somatic bacterial antigen. These antibodies are bactericidal by means of complement-mediated phagocytic mechanisms. Antibodies transferred passively to the fetus across the placenta are considered to be primarily responsible for the low incidence of neonatal meningitis due to *Haemophilus influenzae* in babies under the age of 3 months. Infants less than 18 months of age do not respond to vaccination, and as a result, vaccination before 18 months is not practiced. The vaccine is most effective when given at 24 months. Immunization of pregnant women is not justified.

Hepatitis A

Hepatitis A is excreted in the feces and is acquired through a fecal/oral route. Exposure to the virus is fairly common, and there are no extrahepatic manifestations of disease. Although the pregnant woman may experience more severe symptoms in the third trimester, transplacental viral transfer is unknown. Acute disease during pregnancy may be associated with an increase in abortion rate and prematurity, but direct infection of a neonate is the result of maternal virus shedding at the time of birth. Chronic viral shedding in a prolonged carrier state, however, is not a characteristic of hepatitis A, in contrast to the shedding seen with hepatitis B.

The most important aspect of preventing hepatitis A is hygienic measures. Immune serum globulin may be administered to exposed individuals, but active immunization is not available. Infants born to mothers who are incubating the virus or infants born with acute illness should be treated with immune globulin.

Hepatitis B

This viral illness is characterized by extrahepatic manifestation, a chronic carrier state lasting longer than 6 months, and the long-term sequela of hepatocellular carcinoma. The source of the virus is frequently blood contamination, but it may also be transmitted through sexual contact or vertically from mother to fetus. The risk of the fetus acquiring the virus is 80-90% when the mother has acute infection from the third trimester to the first month postpartum, but only 10-30% when maternal disease is present in the first trimester. Neonatal hepatitis may result from virus acquired from the infant's mother, and maternal disease may predispose to abortion or prematurity. Passive immunization may be practiced by the administration of hepatitis B immunoglobulin as soon after birth as possible in infants born to mothers who have HBsAg (hepatitis B surface antigen), followed by additional doses at the ages of 3 and 6 months. Women seeking obstetric care should be screened for hepatitis B infection; this is especially important in areas where this disease is widespread, such as inner-city populations known to have a high prevalence of intravenous drug use.

Active immunization is available and should be provided to medical personnel and to individuals in high-risk groups, including individuals with multiple sex partners, homosexual and bisexual men, intravenous drug users, individuals being treated for sexually transmitted disease, and prostitutes, as well as residents in correctional facilities or long-term-care facilities. Importantly, vaccination requires three doses over a period of time, something that may be difficult to achieve with some of the population groups mentioned above.

Two forms of active immunization are available for hepatitis B. The older vaccine consists of purified HBsAg from the plasma of known carriers. The same antigen can now be produced by recombinant DNA technology and is the basis for a newer vaccine.

Influenza

Experience from the early part of this century indicated that when influenza occurs in epidemic or pandemic proportions, a more severe effect in pregnant women is noted. Increased morbidity and mortality were noted in the 1918 pandemic, although it is not possible to know whether these effects were due directly to the virus or due to complications of influenza. An increase in the rate of spontaneous abortions has been associated with influenza during pregnancy, but transplacental infection and fetal malformations have not been identified.

A critical aspect in the development of a vaccine for the influenza virus has been the recognition that with each influenza season the natural process of restructuring the major antigenic determinants (hemagglutinin and neuraminidase) makes a different antigenic type of virus emerge as the predominant cause of infection. Currently, these emerging immunotypes are anticipated, and during each influenza season a new vaccine must be issued to provide protection against the prevalent influenza strain. The influenza vaccines currently in use depend on the growth and inactivation of virus particles of the appropriate type. Because the vaccine supplies for use in the population require 6 to 9 months of preparation time and quality assurance testing, an educated prediction must be made as to which influenza strain should be selected for development as a vaccine. Currently, the primary emphasis is to ensure the vaccination of those individuals who are at greatest risk of dying from influenza. This group includes the elderly and those with compromised lung function. Individuals who provide vital community services, including health care workers, police, firefighters, and so on, should be protected from incapacitation by influenza, even though they may not be at risk of death from the illness.

The use of influenza vaccine during pregnancy is not encouraged unless the patient belongs to a group with an underlying disease that may require immunoprophylaxis. There is no particular intrinsic risk to the fetus from the vaccine itself because the virus is inactivated. However, untoward effects of influenza vaccines have been observed, and before undertaking vaccination of a pregnant woman the risk/benefit relationship should be considered.

Measles

Measles vaccines were first introduced in 1963, and several improved vaccine strains have been used since that time. At present vaccination is accomplished with a live attenuated measles virus vaccine that produces immunity in 95% of properly vaccinated individuals, and protective immunity appears to last for at least 20 years. When measles is acquired during pregnancy, it is similar to that in the nonpregnant adult, although morbidity may tend to be somewhat more severe. As with other acute febrile diseases, the illness may interfere with successful completion of the pregnancy, and increases in spontaneous abortion and prematurity have been recorded during epidemics. Rarely transplacental infection occurs, but this has not been definitively linked to any malformation. Vaccination with the live-virus vaccine should not be given during pregnancy. However, children of pregnant women can be vaccinated because there is no evidence that the vaccinated children pose a risk to the pregnant mother. The administration of vaccine may be an appropriate part of postpartum care and presents no known hazard to the newborn.

Recently there has been a spate of adolescents who appear to have inadequate protection despite prior vaccination. Public health campaigns to vaccinate susceptible persons began in 1966. Girls born in that year are now in their prime childbearing years. The acquisition of measles during pregnancy represents a risk primarily to the newborn because the infant may acquire the virus from his or her mother within the first 12 weeks of life and develop measles infection that resembles that seen in older children. In some urban areas young children below the age of 4 years who should have been vaccinated have not received vaccine. In addition, vaccination before the age of 15 months results in inadequate protection. In locales where an inordinate number of preschool cases of measles have been recorded, vaccine is given at both 9 and 15 months of age; the first dose is a monovalent measles vaccine and the second is the polyvalent measles/mumps/rubella vaccine.

Mumps

Routine vaccination against the mumps virus is administered during childhood as a part of the polyvalent measles/mumps/rubella vaccine. The live attenuated virus vaccine has been available since 1968. This vaccine has the ability to infect the placenta, and although no evidence of fetal damage has been demonstrated, the potential for harm precludes its use during pregnancy. The seroconversion rate is very good with the attenuated vaccine and has produced long-lasting immunity similar to that conferred by natural infection. The role of mumps in pregnancy is not entirely clear, but it has been suggested to increase the prevalence of first-trimester abortion and may be related to fibroelastosis in neonates.

Neisseria meningitidis

This organism does not pose any unusual threat to the pregnant woman and is primarily a disease that occurs in small to large outbreaks among individuals in close living situations. Outbreaks have often been seen in young army recruits housed together in a barracks, in young children in day care facilities, or among members of a household. Normally, vaccination is only done in situations where there is a significant risk of disease.

Moreover, the vaccine material, which is capsular polysaccharide, does not cover all serotypes that may potentially cause disease, and there is some question as to whether widespread immunization might select for those low-prevalence strains that are not affected by the currently used vaccine materials.

Because the vaccine material is nonviable purified polysaccharide, it theoretically poses no threat to the fetus if administered during pregnancy. However, its safety profile during pregnancy is unknown. Therefore, it should not be given unless specifically required by the occurrence of an epidemic.

Plague

Fortunately, *Yersinia pestis* is a rare cause of infection because many of the environmental conditions related to previous worldwide epidemics have been altered. However, its morbidity and mortality are striking, and despite the rarity of the disease, it still exists. Areas of endemicity in the United States include the Southwest. The extraordinary virulence of this organism has been noted, but effects of this disease beyond those seen in nonpregnant individuals have not been documented in pregnant women. Treatment for plague victims involves antibiotic therapy, and immunization is only rarely practiced. A formalinized preparation of bacteria is the immunogen, and this is only given to individuals who work with the organism in a laboratory setting or individuals who must do field work in areas where they will be in contact with animals that may be infected with the plague.

Poliomyelitis

This enterovirus has been responsible for significant epidemics of morbidity and mortality in the not-so-distant past. In the prevaccine era the effects of disease when acquired during pregnancy appeared to be more severe, and damage to the fetus due to anoxia was commonly observed. The polio virus can be transmitted across the placenta near the time of parturition or may enter the fetus by ingestion during delivery. Maternal antibody crosses the placenta and provides protection to the fetus. When the disease is acquired by the mother several weeks before the infant's encounter with the virus, the infant may be protected. If the mother acquires polio near the time of delivery, there is little opportunity for the synthesis and transplacental transport of protective antibody.

Vaccines, including the inactivated Salk vaccine that became available in 1955 and the attenuated live Sabin vaccine that became available in 1961, have been responsible for vast reductions in morbidity and mortality. For primary immunization in the United States the oral vaccine is considered the standard, although there is currently some controversy about the need to shift to inactivated vaccines that do not have the potential of causing paralytic disease. Clearly, for pregnant women, only the inactivated vaccine can be recommended if the risk warrants immunization.

Rabies

The rabies virus is found in animal reservoirs, and when acquired by humans through animal bites, it is uniformly fatal unless adequate intervention is undertaken. Vaccination of domestic animals is a very important aspect of the control of this disease. The immunoprophylactic strategy in exposed humans involves the administration of human rabies immunoglobulin and active vaccination with inactivated rabies virus. Currently, a human diploid-cell-grown inactivated virus is available, obviating some of the problems related to the use of the older duck embryo vaccine. If a woman is exposed to rabies virus during pregnancy, she should undergo the same treatment as a nonpregnant individual.

Rubella

The devastating effects of transplacentally acquired rubella on the fetus are well known to obstetricians and will not be recounted here. Our current understanding of the hazards of transplacental infections has largely come from the lessons learned from rubella acquired in pregnancy. As a part of routine childhood immunizations, a live attenuated virus vaccine is administered and should provide protection until women reach childbearing years. However, the prevalence of childhood immunization is not 100%, and antibody titers can decline in some individuals to levels that fail to provide protection. Reinfections or infection after immunization may be related to diminished secretory IgA (the upper respiratory tract is the portal of entry for the virus), but a secondary immune response will probably be sufficient to prevent fetal damage. The immunologic status of women presenting for prenatal care is routinely evaluated. For those who are seronegative, postpartum vaccination is practiced. Although some 250 pregnant women have inadvertently been exposed to vaccine, no adverse fetal effects have been demonstrated, although the vaccine virus is able to cross the placenta. Vaccination during pregnancy remains contraindicated.

Streptococcus pneumoniae

The pneumococcus causes pneumonia and meningitis. Infection with *Streptococcus pneumoniae* is more common in adults than children and when accompanied by bacteremia has a mortality rate of approximately 25%. Because there are more than 80 serologic types based on differences in the polysaccharide capsule, the development of a vaccine once appeared impossible. However, because 90% of the infections are caused by 23 pneumococcal serotypes, a polyvalent vaccine consisting of capsular polysaccharide material from each of these 23 pneumococcal types has proved to be a satisfactory vaccine material. The indications for vaccination are risk of pneumococcal infection and ability to respond to the immunogen. Severely immunocompromised hosts and infants under 2 years of age are not candidates for vaccine. The elderly, especially those in long-term-care facilities, benefit from the vaccine. Most women of childbearing age are not among the groups typically vaccinated. Therefore, the indications for vaccination are the same as those for other individuals. The effect of the vaccine on the fetus is unknown.

Tetanus

Like diphtheria, tetanus prophylaxis depends on active vaccination with a toxoid vaccine. The toxoid may be used in pregnancy, and protective antibodies are known to cross the placenta. Because the mortality associated with tetanus may be as high as 60%, tetanus immunity may be boosted as part of prenatal care if no booster has been administered within 10 years. Likewise, individuals with no history of primary immunization should undergo a standard immunization protocol.

Tuberculosis

Immunization against *Mycobacterium tuberculosis* is not practiced in the United States. Tuberculosis may adversely affect the neonate. The organism may interfere with placental function, although transplacental infection of the neonate is rare. However, the infant may acquire the organism by aspiration and become infected, with the lungs as the primary site of infection. Therapy of the infection is the preferred method of controlling the disease in the United States. It is important to remember that emigrés may have had bacillus Calmette-Guérin live attenuated vaccine and will display a profound immune response to the use of skin-testing materials.

Tularemia

Tularemia is a disease resembling plague and is mainly caused by contact with infected animals. Adequate prevention is achieved by avoiding contact with the animal sources of infection. For laboratory personnel who may come in contact with *Francisella tularensis*, as well as sheepherders, sheepshearers, and trappers, a live attenuated vaccine is available through the CDC. The development of immunity is rather slow and may require two doses of vaccine at 2-month intervals.

Varicella

There is no active vaccine for varicella, and the majority of adults have immunity to the virus because of infection during childhood. The virus is quite contagious, and when varicella is acquired during childhood, few cases are attended by serious complications. During pregnancy, however, varicella may cause a severe pneumonia. Vertical transmission is possible during the first trimester and may produce a variety of effects in the fetus.

Infection in newborns is accompanied by mortality of approximately 35%. Although a live attenuated vaccine is available, it is only used in immunosuppressed children. Other prophylactic efforts have emphasized the use of zoster immunoglobulin for passive protection. Immune serum is not ordinarily given to pregnant women exposed to the virus, but it is administered to infants born to mothers who acquire the virus within 4 days of delivery or 2 days after delivery. Immunoglobulin should be given within 72 hours of exposure to the virus to achieve the desired effect.

Yellow Fever

Yellow fever is a highly virulent mosquito-borne viral disease that results in fever, jaundice, hemorrhage, and hypovolemic shock. Its virulence is not known to be affected by pregnancy. Because it is a disease of the tropics, only certain travelers are in need of the live attenuated virus vaccine. The vaccine should not be used in pregnancy if it can be avoided by a change in travel plans. If travel is unavoidable, vaccination must be undertaken because of the rapid onset and high mortality associated with this disease.

Bibliography

Albritton WL. Biology of *Haemophilus ducreyi*. Microbiol Rev 1989;53:377-389

Baker CJ, Noya FJ. Potential use of intravenous immune globulin for group B streptococcal infection. Rev Infect Dis 1990;12:476s-482s

Baker CJ, Rench MA, Kasper DL. Response to type III polysaccharide in women whose infants have had invasive group B streptococcal infection. N Engl J Med 1990;322:1857-1860

Baughn RE. Role of fibronectin in the pathogenesis of syphilis. Rev Infect Dis 1987;9:372s-385s

Chesney PJ. Clinical aspects and spectrum of illness of toxic shock syndrome: overview. Rev Infect Dis 1989;11:1s-7s

Clark RA. The human neutrophil respiratory burst oxidase. J Infect Dis 1990;161:1140-1147

Cohen MS, Britigan BE, Hassett DJ, Rosen GM. Phagocytes, O_2 reduction, and hydroxyl radical. Rev Infect Dis 1988;10:1088-1096

Elliot B, Brunham RC, Laga M, Piot P, Ndinya-Achola JO, Maitha G, et al. Maternal gonococcal infection as a preventable risk factor for low birth weight. J Infect Dis 1990;161:531-536

Hofstad T. Current taxonomy of medically important nonsporing anaerobes. Rev Infect Dis 1990;12:122s-126s

Johnson RE, Nahmias AJ, Magder LS, Lee FK, Brooks CA, Snowden CB. A seroepidemiologic survey of the prevalence of herpes simplex virus type 2 infection in the United States. N Engl J Med 1989;321:7-12

Johnston MM, Sanchez-Ramos L, Vaughn AJ, Todd MW, Benrubi GI. Antibiotic therapy in preterm premature rupture of the membranes: a randomized prospective, double-blind trial. Am J Obstet Gynecol 1990;163:743-747

McGregor JA, French JI, Richter R, Franco-Buff A, Johnson A, Hillier S, et al. Antenatal microbiologic and maternal risk factors associated with prematurity. Am J Obstet Gynecol 1990;163:1465-1473

Minkoff HM, ed. HIV disease in pregnancy. Obstet Gynecol Clin N Amer 1990;17:489-668

1989 Sexually transmitted diseases treatment guidelines. Rev Infect Dis 1990;12:577s-690s

Pastorek JG II, ed. Sexually transmitted diseases. Obstet Gynecol Clin N Amer 1989;16:453-702

Schlievert PM. Role of toxic shock syndrome toxin 1 in toxic shock syndrome: overview. Rev Infect Dis 1989;11:107s-109s

Stamm WE, Hooton TM, Johnson JR, Johnson C, Stapleton A, Roberts PL, et al. Urinary tract infections: from pathogenesis to treatment. J Infect Dis 1989;159:400-406

Stevens JG. Human herpes viruses: a consideration of the latent state. Microbiol Rev 1989;53:318-332

van den Broek PJ. Antimicrobial drugs, microorganisms and phagocytes. Rev Infect Dis 1989;11:213-245

Yonekura ML. Risk factors for postcesarean endomyometritis. Am J Med 1985; 78(suppl 6B):177-187

zur Hausen H, Schneider A. The role of papillomaviruses in human anogenital cancer. In: Salzman NP, Howley PM, eds. The papovaviridae. Vol 2. New York: Plenum Press, 1987:245-263

Index